Methanol Fuel in Transportation Sector and Fuel Cells

Edited by Lindiwe Khotseng and Sello Ntalane Seroka

Published in London, United Kingdom

Methanol Fuel in Transportation Sector and Fuel Cells
http://dx.doi.org/10.5772/intechopen.111074
Edited by Lindiwe Khotseng and Sello Ntalane Seroka

Contributors
Adebare Nurudeen Adewumi, Ahsan Hanif, Fehrs Adu-Gyamfi, Huaneng Su, Joseph Parbey, Lindiwe Khotseng, Mahir Es-Saheb, Marcello Romagnoli, Michael Gyan, Muhammad Kashif Jamil, Muhammad Mujtaba Abbas, Muhammad Usman, Sabejeje Akindeji Jerome, Sello Ntalane Seroka, Veronica Testa, Yasser Fouad, Yi Cheng Yi

First published in London, United Kingdom, 2024 by IntechOpen
IntechOpen is the global imprint of INTECHOPEN LIMITED, registered in England and Wales, registration number: 11086078, 167-169 Great Portland Street, London, W1W 5PF, United Kingdom

British Library Cataloguing-in-Publication Data
A catalogue record for this book is available from the British Library

Additional hard and PDF copies can be obtained from orders@intechopen.com

Methanol Fuel in Transportation Sector and Fuel Cells
Edited by Lindiwe Khotseng and Sello Ntalane Seroka
p. cm.
Print ISBN 978-0-85466-110-7
Online ISBN 978-0-85466-109-1
eBook (PDF) ISBN 978-0-85466-111-4

Meet the editors

Prof. Lindiwe Khotseng is a professor in the Chemistry Department, University of the Western Cape, South Africa. Her research focus is on energy nanomaterials. She is a rated researcher in South Africa with published works on fuel cell and biomass waste in international journals and specialized books. She has supervised MSc and Ph.D. students and is the recipient of the 2013 Outstanding Black Researcher award from the Department of Trade and Industry, South Africa.

Ntalane Sello obtained his Ph.D. in Chemistry from the University of the Western Cape, South Africa, in 2023. Currently, he is a postdoctoral researcher at the University of the Western Cape-CSIR. Dr. Sello conducts research on renewable energy technologies. His research focuses on the development of nanomaterials for biomass conversion, energy conversion, and energy storage. He employs nanotechnology as a catalyst for the production of nanostructured solar cell systems based on silicon thin films. He also studies waste-to-energy conversion and conversion of biomass waste to biofuels and biomaterials.

Contents

Preface

Climate change and the unsustainability of fossil fuels are the main reasons for the increasing demand for renewable energy sources. Methanol can be produced from a wide variety of renewable feedstocks, the most significant being biomass, making it one of several alternative energy sources available today. Methanol is also a clean-burning, biodegradable fuel, making it environmentally friendly.

Methanol use in transportation fuel is growing due to its efficient combustion, ease of distribution, and wide availability around the world. Methanol is used as an alternative fuel for compatible vehicles, with more research focusing on its use as a sustainable marine fuel. Methanol is also used as a fuel in fuel cells, which is one of the technologies that plays a key role on the road to a greener future.

The first chapter discusses the major challenges and solutions in introducing methanol as a fuel in the transportation sector. These challenges include cold start issues, enhanced NOx emissions, 100% methanol utilization, and lubricant-oil deterioration. The chapter also discusses different techniques that can be used to optimize and predict the use of methanol in internal combustion engines.

The subsequent chapters discuss progress on alternative nanomaterials for cathode electrodes in direct methanol fuel cells, improved performance in direct methanol fuel cells due to nanostructured carbon support materials with improved physical and chemical properties, as well as principles and performance of direct methanol fuel cells.

Lindiwe Khotseng
Professor,
Department of Chemistry,
University of the Western Cape,
Cape Town, South Africa

Dr. Sello Ntalane Seroka
Department of Chemistry,
University of the Western Cape,
Cape Town, South Africa

Chapter 1

Introductory Chapter: Methanol Fuel – New Developments and Applications

Lindiwe Khotseng and Sello Ntalane Seroka

1. Introduction

Climate change is a major threat that the world is dealing with today, and more efforts are needed in all economic sectors to reduce greenhouse gas (GHG) emissions. Especially in the transport sector, thus fossil-free fuels are required to reduce the negative impact caused by climate change. The most effective fuels are those with minimum GHG emissions and minimum pollutants, while compatible with common internal combustion engines and fuel infrastructure. There are various alternative fuel options including methanol, ethanol, and ammonia [1]. Comparing the three fuels, methanol has the advantage of being similar to traditional fuels. It is a hydrocarbon and a liquid. With stronger commitments to address climate change, the focus is on methanol as one of the fuels to be considered, especially for maritime engines, as it is making it easier to allow dual-fuel engines to be used in the marine sector [2]. Methanol is also used as fuel in other applications including fuel cells and the motor industry. This book reports on various developments in using methanol as a fuel for transportation and direct methanol fuel cells.

Methanol is a clear, colorless, neutral, and polar liquid at ambient conditions, with boiling and freezing points of 64.6°C and −97.6°C, respectively [3]. Methanol is produced using fossil fuels (i.e., natural gas, coal, and petrochemicals); it can also be produced from renewable sources such as biomass raw materials, agricultural raw materials (i.e., sugar cane, cereals, and rice), timber, and urban waste. More research is focusing on producing methanol from renewable sources as its production from fossil fuels results in a greenhouse emission. The growing demand for green methanol in the maritime industry has led to a growing number of project announcements for green methanol around the globe. It is predicted that the green methanol market will increase by 39.6%, which will have a significant improvement in climate change (**Figure 1**) [4].

2. Methanol in the transport sector

Methanol possesses some interesting properties that make it more performing compared to conventional fuels for internal combustion engines, including high latent heat, fast-burning velocity, no carbon-to-carbon bonds, and high octane rating, and

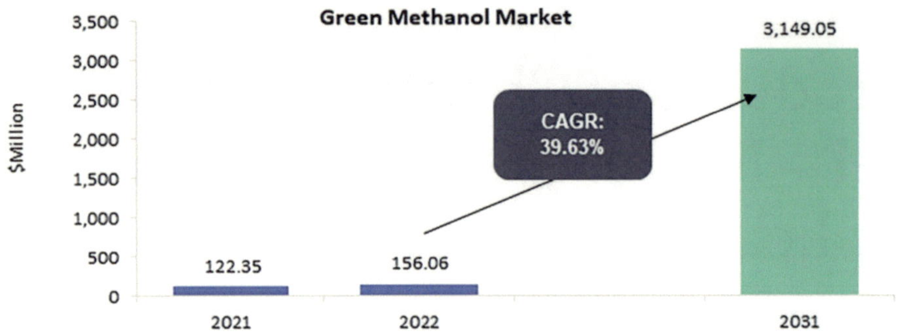

Figure 1.
Green Methanol Market, $Million, 2021, 2022, and 2031 [4].

hence higher compression ratios and higher knock resistance for an increased engine efficiency [5].

Moreover, methanol is a fuel of choice due to its clean-burning and less explosive nature, which results in less gas emission (GHG) compared to other fuels, which has a significant effect on bettering the environment. When used as fuel transportation, it can be used straight for a gasoline component called Methyl tert-butyl ether (MTBE), as a blending component of fuel, or for the production of fuel components including a diesel component called fatty acid methyl ester FAME [4]. Countries like China and the United States use methanol-gasoline blends due to high demand for fuel and also its benefit in reducing GHG emissions impact and its use has been increasing in other regions like Asia and Europe [5].

Recently, more research focused on using methanol in the marine sector has grown rapidly, leading to commitments from some of the world's largest shipping companies due to several advantages including being the lowest-cost carbon-neutral shipping fuel, by total cost of ownership (TCO), across a wide range of vessels and applications when compared to a suit of fuels including ammonia, liquefied biogas, electricity and hydrogen [6].

3. Methanol in fuel cells

Methanol is used as a fuel in direct methanol fuel cells, which is a variant of proton exchange membrane fuel cells where the mixture of methanol and water is used instead of hydrogen and indirectly via steam reforming into a hydrogen-rich gas mixture in HT-PEMFCs. Methanol is used as a fuel in direct methanol due to several advantages including its easy handling as a liquid mixture compared to the hydrogen storage required for low-temperature fuel cells or the added reforming system needed in HT-PEMFC. Methanol as fuel in fuel cells still faces challenges like the costs due to the Pt-based catalysts used, methanol crossover, low efficiency of below 30%, and poor electrode kinetics [7].

Methanol in fuel cells converts the liquid fuel, i.e. methanol into electricity in combination with oxygen producing only waste heat, water vapour and small amount of carbon dioxide [8].

$$\text{Anode}: CH_3OH + H_2O \rightarrow 6H^+ + 6e^- + CO_2 \text{ (oxidation)} \quad (1)$$

$$\text{Cathode}: \frac{3}{2}O_2 + 6H^+ + 6e^- \rightarrow 3H_2 \text{ (reduction)} \tag{2}$$

$$\text{Overall reaction}: CH_3OH + \frac{3}{2}O_2 \rightarrow 2H_2O + CO_2 \text{ (redox reaction)} \tag{3}$$

Pt is used as a catalyst for both half reactions. This contributes to the loss of the cell voltage potential, as any methanol that is present in the cathode chamber will oxidize. If another catalyst could be found for the reduction of oxygen, the problem of methanol crossover would likely be significantly lessened [9]. It is necessary to develop other low-cost, durable, safe and efficient alternatives to replace Pt catalysts in DMFC [10].

Author details

Lindiwe Khotseng* and Sello Ntalane Seroka
Department of Chemistry, University of the Western Cape, Bellville, Cape Town, South Africa

*Address all correspondence to: lkhotseng@uwc.ac.za

References

[1] AMF TCP work on methanol 2018-2020. Available from: https://www.iea-amf.org/content/fuel_information/methanol.html

[2] MAN Energy Solutions. The Benefits of Methanol. 2023. Available from: https://www.man-es.com

[3] Offermanns H, Schulz K, Brandes E, Schendler T. Methanol utilisation technologies. In: Bertau M, Offermanns H, Plass L, Schmidt F, Wernicke HJ, editors. Methanol: The Basic Chemical and Energy Feedstock of the Future: A singer's Vision Today. Berlin/Heidelberg, Germany: Springer; 2014. pp. 327-601

[4] BIS Research. Green Methanol Market. 2022. Available from: https://bisresearch.com/industry-report/green-methanol-market.html

[5] Araya SS, Liso V, Xiaoti CX, Li N, Zhu J, Sahlin SL, et al. A review of the methanol economy: The fuel cell route. Energies. 2020;**13**(3):596. DOI: 10.3390/en13030596

[6] IRENA. Innovation Outlook Renewable Methanol. 2021. Available from: https://www.irena.org

[7] Visser ED, Seroka NS, Khotseng L. Catalytic properties of biochar as support material potential for direct methanol fuel cell: A review. ACS Omega. 2023;**8**(44):40972-40981. DOI: 10.1021/acsomega.3c02283

[8] Methanol Institute. Renewable Methanol Report. Dec 2018. Available from: https://www.methanol.org/wpcontent/uploads/2019/01/MethanolReport.pdf

[9] Direct Methanol Fuel Cell. Available from: https://en.wikipedia.org/wiki/Direct_methanol_fuel_cell

[10] Qin C, Tian S, Wang W, Jiang Z-J, Jiang Z. Advances in platinum-based and platinum-free oxygen reduction reaction catalysts for cathodes in direct methanol fuel cells: A review. Frontiers in Chemistry. 2022;**10**. DOI: 10.3389/fchem.2022.1073566

Chapter 2

Core Challenges and Prospects of Methanol Utilization, Prediction and Optimization for Sustainable Environment

Muhammad Usman, Muhammad Kashif Jamil, Ahsan Hanif, Muhammad Mujtaba Abbas, Mahir Es-Saheb and Yasser Fouad

Abstract

The transportation sector stands at the forefront of global challenges, where its significant contribution to greenhouse gas emissions and air pollution has become an urgent matter demanding immediate attention. For addressing these compelling concerns and leading the automotive industry toward a sustainable future, it is, therefore, imperative to explore the realm of alternative fuel that can effectively mitigate the environmental impact of automobiles. Methanol, a renewable alternative fuel, has gathered quite an attention due to its potential to be used as a wonderful alternative to neat gasoline in spark ignition engines. However, there are some core challenges that must be addressed to utilize methanol on a commercial scale in the transport sector. These core challenges include cold start issues, enhanced NOx emissions, 100% methanol utilization, transportation concerns and lubricant oil deterioration. In this chapter, these challenges along with their potential solutions have been discussed in detail. Moreover, different techniques such as artificial neural network and response surface methodology have been discussed to predict and optimize the usage of methanol in SI engines. The adoption of methanol, as an alternative to gasoline, will help us achieve some important sustainable development goals, thus fulfilling the promise of a sustainable future for the upcoming world.

Keywords: transportation industry, greenhouse gases, sustainable environment, alternative fuels, responsible consumption, optimization, prediction

1. Introduction

The desire for a luxurious lifestyle has forced mankind to deploy energy resources at a very fast rate. With each passing day, fossil fuels are getting consumed immensely to fulfill this particular desire. Non-renewable fossil fuels are being majorly employed in two sectors: Power Generation and Transportation [1]. In 2017, it was reported that the global crude oil production had reached approximately 4700 million tons

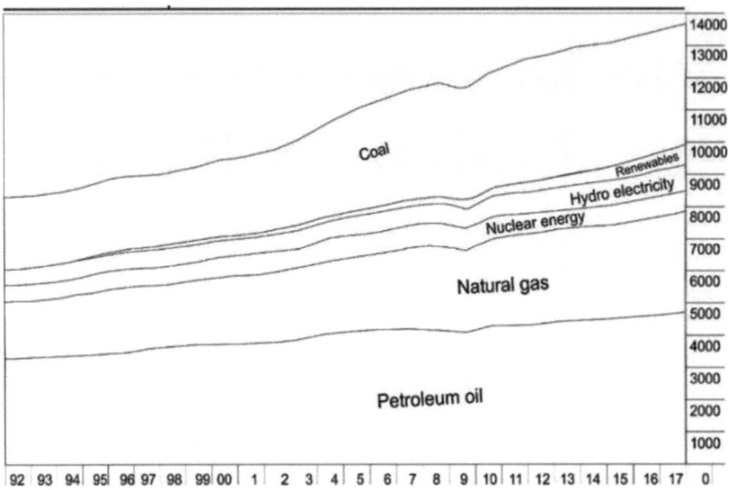

Figure 1.
World energy consumption (y-axis) in million tons.

as shown in **Figure 1**. The transportation sector performs a significant role in the socioeconomic development of any country around the world. Currently, the primary concerns for transportation sector include the oscillation in the fuel prices, vehicular exhaust emissions and constantly depleting fossil fuels [2].

As per reports of EIA, in terms of years, the predicted lifespan of fossil fuel reserves left is only about 50.7 years [3]. Moreover, the present oil reserves of world approximated about 1342 billion barrels in the year 2009 [4]. Moreover, the EIA has also notified that the worldwide oil production in the year 2007 reached approximately 85 million barrels [5]. Dr. Colin Campbell, an American biochemist, proposed that the total fuel reserves of world will be exhausted in upcoming four decades if the oil production stays at present rate. The unpredictable fluctuation in fuel prices has also become a major concern for both developed and developing countries [6]. The main reason for this sudden fluctuation in fuel prices is the continuous depletion of fossil fuels along with the increase in world population and the corresponding energy demand [7]. Most of the developing countries are spending a large proportion of their budgets for importing the petroleum products to meet the ongoing energy demand that greatly affects the economic stability and social development of these countries [8].

The immense usage of fossil fuels is not only responsible for its depletion but also results in various socioenvironmental impacts including global warming, tempered air quality, extreme weather events, ecological disruptions, traffic congestion and noise pollution [9]. The major sources of greenhouse gases (GHG) are industries and the constant increase in the usage of comforting vehicles. It has been proposed that by the year 2050, the number of vehicles including busses, cars and trucks will increase to approximately 2.5 billion if the production and use of these particular automotive continue at the constantly increasing rate [10]. This makes the automotive sector the largest contributor of the total GHG emissions which ranges up to 60% of the total emissions. According to EIA, total global energy consumption is projected to increase by approximately 56% by the year 2040 in contrast with year 2010 [11]. This alarming increase in energy consumption indicates an astounding amount of greenhouse gases

(GHGs) emissions, comprising at least 80% of carbon emissions being discharged into the atmosphere causing environmental deterioration and abrupt climate change.

The consequences of the combustion of fossil fuels include global warming resulting due to considerable emissions of CO2, NOx and mainly SOx, reported deadly for mankind and wildlife. It was reported, by IEA in 2015, that the total final consumption of coal from 1973 to 2014 increased from 631 MTOE to 1075 MTOE [11], as shown in **Figure 2 (a)** and **(b)**. Shares of world coal consumption in various sectors in 1973 and 2013 can be predicted from the given pie charts.

On the other hand, sector wise consumption of oil had reached an amount of 2252 MTOE by the end of 1973 and by the end of 2013, this sector wise oil consumption had reached an amount of 3716 MTOE which can be observed in **Figure 3**.

The overall sector wise consumption of natural gas in 1973 and 2013 was 652 MTOE and 1401 MTOE, respectively, which can be seen in **Figure 4(a)** and **(b)**. According to a review by the World Bank in year 2000, these deleterious gases have resulted in more than 70,000 deaths annually [12]. Several nitrous and sulfur oxides are the key emissions of the road-based transport sector. These gases are not just irritant but also responsible for acidic rain which has hazardous impacts on the crops as well as the aquatic life. The major cause of sulfur oxide emissions is the combustion of coal in the industrial sector for the purpose of power generation. One-third of the

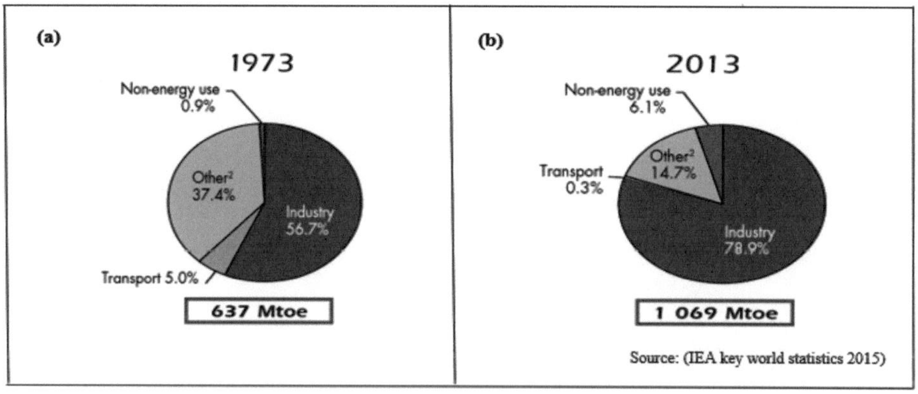

Figure 2.
(a) and (b). Coal consumption in 1973 and 2013.

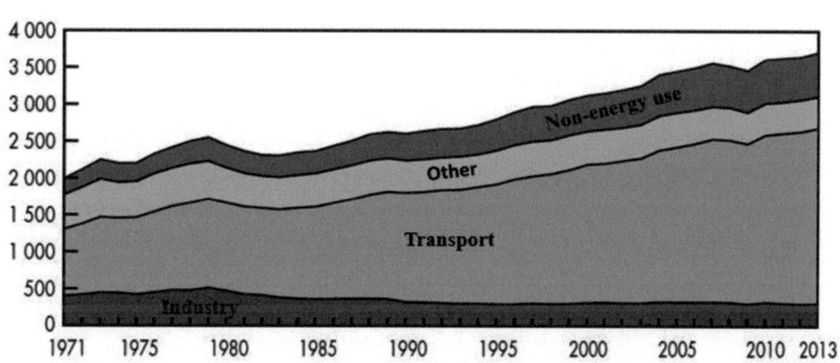

Figure 3.
Overall sector wise oil consumption in MTOE from the year 1971 to 2013.

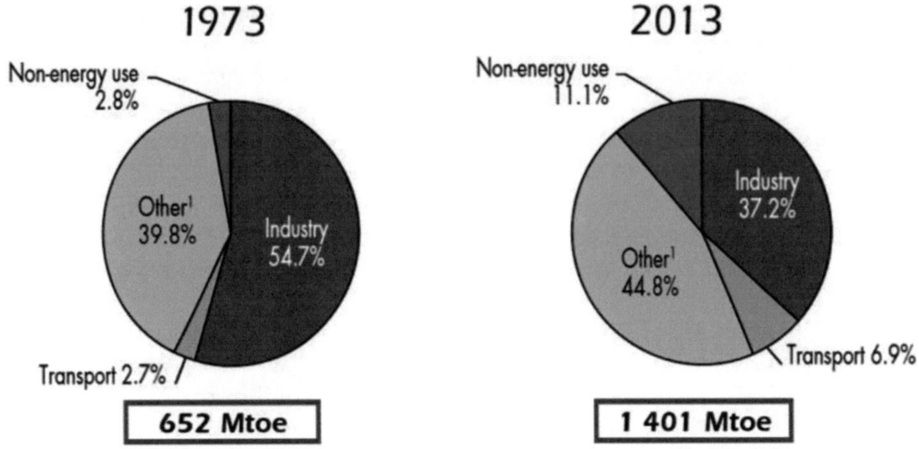

Figure 4.
(a) and (b). Natural gas consumption in 1973 and 2013.

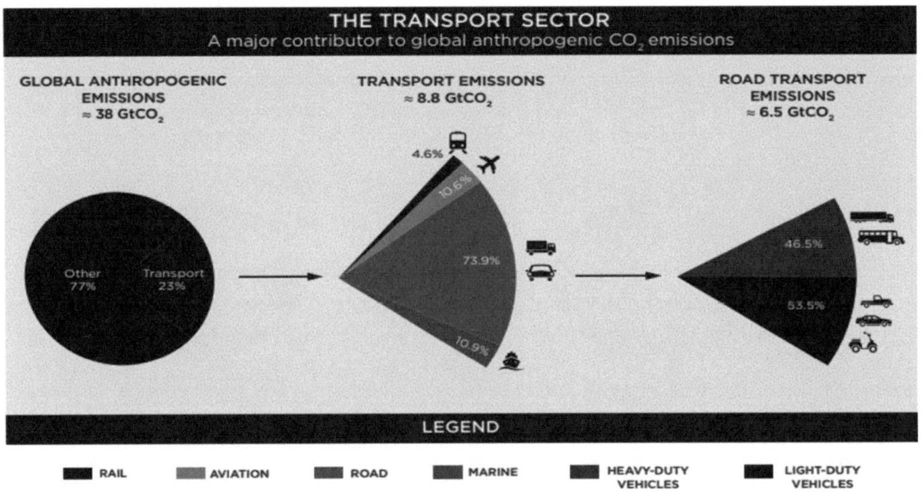

Figure 5.
Global anthropogenic CO_2 emissions from the transport sector.

total carbon emissions are also directly linked to the industrial division. These carbon emissions are responsible for the increase in the average temperature of earth. It has been reported *via* a survey that global earth temperature has been amplified at the rate of 0.6 degrees Celsius. The USA Department of Energy predicted that the earth's average temperature would be expected to increase 1.7–4.9 degree Celsius by the year 2015 due to 54% increased carbon emissions when compared with the year 1990 [13].

The transport sector has served as a major contributor to the increase in CO_2 emissions which resulted in global warming. As per **Figure 5**, it can be seen that transport sector contributes 23% global anthropogenic CO_2 emissions out of which about 73.9% of the total carbon emissions in this sector is mainly due to road-based vehicles [14].

The dwindling of fossil fuels around the globe has been a hot topic of discussion in world monetary and political circles in preceding decades, especially in developed consumer nations like Europe and America. The central focus of this conversation revolves around alternative sources of energy once the fossil fuel reserves are depleted [15]. Undoubtedly, the key variables in formulating and making informed decisions regarding alternative energy sources lie in their financial viability and how they can be effectively utilized to meet the needs and preferences of consumers. During recent decades, the energy mix of consumer countries shifted toward renewable resources like hydro, nuclear, biomass, wind and solar instead being dependent on conventional fuels [16]. An immense effort has been incorporated for the advancement of renewable technologies but still, these technologies have not been commercialized on massive scale because the calorific value or energy content in petrol or diesel cannot be matched by these renewable energy sources and high initial capital investment is a major barrier toward the development of these renewable sources as complete alternative to fossil fuels. But still the modern area of research is to improve the effectiveness of turbomachinery and to look at alternatives of fossil fuels which are not only renewable in nature but can also be easily available so that they can be produced on small scale easily, therefore resulting in reduction of dependence on fossil fuels [17]. Therefore, there is need to look for renewable fuel sources which can not only reduce emissions but also lessen reliance on fossil fuels. If the utilization of fossil fuels is continued at a present rate, no sincere effort will be made to explore new oil reserves or fuel alternatives, then there will be a time when all the present fossil fuel reserves will drain out and no more fuel resources will be left to generate energy and run the engines [18].

The basic operating principle of automobiles involves the burning of fuel to produce heat energy that is subsequently converted into mechanical energy for the propulsion of vehicle. Consequently, the emissions get exhaled into the atmosphere and deteriorate the environment so badly. Spark Ignition (SI) engines operate on rich conditions, and they generate more carbon emissions when compared to the diesel engines [19]. SI engines are widely used in automotive, lawn mowers, portable small generators and other low power applications due to their lower maintenance cost and lower initial investment. Whereas the substantial depletion of fossil fuel resources, alarming environmental risks and fluctuations in fuel costs are the paramount concerns. Developed countries have two challenges, first, they are releasing more carbon emissions into the air, causing climate change and secondly, they are running out of fossil fuels like oil and coal [20]. The modern area of research is focused on looking for green fuel sources which are not only compatible to this particular type of engines but also produce less emissions and give higher thermal efficiencies, so that they can be served as good alternative of fossil fuels. Alcohol is considered renewable in nature and presents excess oxygen which plays a significant role in reducing emissions and ensuring complete combustion with minimal losses, when compared to gasoline. Among many members in the alcohol family, methanol and ethanol proved to be more desirable due to affordability and reduced emissions [21].

Expansion of cities and advancements in industrial segments prompted environmental concerns. Petroleum-based products play a pivotal role in meeting the energy needs across various sectors crucial for a country's development, including industrial, power and transport segments. But emissions from the combustion of such fuels promote environmental issues [22]. A drawn-out resolution for these issues is required for a maintainable turn of events.

In this manner, sustainable power sources are emerging as productive and successful resolutions for ecological issues [23]. Bioenergy from biomass has been attracting countries in recent years to attain benefits from bioenergy and utilize this energy as substitute of these fossil fuels. The major beneficial outcomes of bioenergy are listed below [24].

- Reduction in emissions
- Renewable in nature
- Responsible for economic growth
- Enhances agriculture
- Diminishes landfills
- Energy security
- Decreases necessity on fossil fuels
- Many products can be produced like ethanol and methanol from bioenergy

2. Overview of methanol

Methanol, CH_3OH, is the simplest hydrocarbon with oxygen. It stands among the five most widely used chemicals in the world [25]. In the past, methanol was mainly used for the production of adhesives, paints, silicones, LCD screens and pharmaceutical products. It has also been used in the wood and automotive industry on commercial scale. During recent years, methanol has been largely employed for power purposes, with approximately 0.020 billion tonnes of methanol being generated annually [26]. Its major uses, which are constantly increasing day by day, include its usage as a fuel as well as a fuel blend component.

One prime reason to investigate methanol as a transportation fuel is its scalability. The simplicity in its making and a wider choice of raw feedstocks presents methanol a convincing entrant for a sustainable alternative energy source with the ability to immensely diminish the carbon footprint of fossil fuels used in transportation sector [27]. The better efficiency with which it can be produced as compared to its fellow alcohols (ethanol, propanol and butanol), coupled with the increased brake thermal efficiency it generates in the internal combustion engines, means that it generates a combined impact on enhancing primary power utilization. Although methanol produces hydrocarbon emissions at the same level to gasoline, its particulate matter contents are substantially lower when compared to the complex synthetic hydrocarbon fuels [28]. The reason for this is its better burning attributes and mono carbon in a molecule by nature. Thus, it can be concluded that methanol has certain potential advantages when it comes to energy security, sustainability and air quality [29].

Methanol is majorly extracted from reforming of natural gas with steam, which transforms this gaseous mixture into methanol through distillation under high temperature and pressure circumstances in manifestation of nickel catalyst [30]. In consequence, methanol is obtained as organic and biodegradable liquid. Another way to produce methanol is to allow the reaction between coal and oxygen in gasifier for generation of synthesis gas incorporating, H_2, CO, CO_2 and minute traces of inert gases like N_2, Ar, CH_3. The reaction in gasifier served as reason to generate syngas, which is the blend of CO, CO_2 and H_2 under conditions of 50–100 bar pressure and

250–300°C temperature, employing copper and zinc-based catalyst to speed up the reaction. These catalysts remain active at 200°C temperature and selective for the production of H_2 and CO_2 [31].

Methanol is extracted from the synergetic response of CO, CO_2 and H_2 in the manifestation of catalyst. The combination of the gases mentioned above is known to be synthesis gas which is usually extracted from gasification of biomass at high temperature and pressure [32]. Then, Syngas is treated to remove impurities like tar and to adjust the carbon-to-hydrogen ratio. Methanol can also be produced from pyrolysis of wood. Biomass can serve as good option in production of methanol. As Pakistan is an agricultural country and has lot of biomass potential, it should be harnessed in order to provide relief to economy by reducing imports of fossil fuels and by increasing production of methanol through biomass [33]. When the production cost of methanol is estimated in terms of equivalence energy and compared to the production cost of gasoline, then still production cost of methanol is lower as compared to gasoline production cost. The basic alcohol can be created by the hydrogenation of carbon monoxide, the natural gas steam reforming, destructive refinement of biomass and wood [34].

There are numerous desirable attributes of methanol which make it an exceptional fuel for spark ignition engines. These include,

- Lower theoretical air-fuel ratio
- Higher latent heat
- Higher specific energy per unit of fuel-air mixture
- Lower combustion temperature
- Higher hydrogen-to-carbon ratio

Out of all these properties, the most important property of methanol is that it can provide clean ignition. It is used in the automotive sector as an alternative to gasoline for air quality goals. With a single carbon atom, methanol cannot easily produce the carbonaceous particulate matter which is common for long-chair hydrocarbon fuels [35]. Another advantage is lesser vehicular tailpipe pollutants when energized with gasoline-methanol mixtures as methanol carries a low boiling point, that is, 65°C which aids burning of fuel within the expansion stroke of the engine working cycle. Complete combustion is responsible for lower hydrocarbon emissions. The more oxygen percentage (50 percent by volume) of methanol along with its simpler composition led to less carbon monoxide emissions. These characteristics increase its significance to be utilized as substitute to gasoline fuel or additive to gasoline fuel in spark ignition engines [36]. Furthermore, methanol possesses inferior vapor pressure and higher latent heat, which confines its usage as fuel in automotive owing to cold start issues in winter. This cold start problem could be resolved by utilizing mixtures of methanol and gasoline. Numerous strategies have been implemented like blended fuel, reforming and heating of fuel along with induction air heating to use methanol as alternate to fossil fuels [37]. Interestingly, these fuel bends having various proportions of methanol in gasoline fuel allow SI engine to run at more compression ratios due to increased octane number [38].

2.1 Physical and chemical properties of methanol

In this section, the physical and chemical properties of methanol are discussed in detail. Furthermore, these properties will be linked to expected consequences of engine design and engine's performance and emission parameters. **Tables 1** and **2** present the physical and chemical properties of methanol and gasoline. A qualitative analysis will be performed, in this section, between these two fuels in order to identify the reliability of methanol as a sustainable alternative fuel.

3. Core challenges in methanol utilization

As the search for sustainable alternative to conventional fuels continues, methanol has emerged as one of the potential solutions for internal combustion engines, especially spark ignition engines. However, there are several challenges that need to be addressed before the deployment of methanol fuel as an alternative to neat gasoline in spark ignition engines. These challenges include cold start issues, lubricant oil deterioration, increased NOx emissions, 100% methanol usage and transportation

Property	Gasoline	Methanol
Vapor density (STP) [kg/m^3]	388	142
Dynamic viscosity (20°C) [mPas]	0.56	0.61
Surface tension (20°C) [mN/m]	21.6	22.1
Heat of vaporization [kJ/kg]	190–350	1100
Boiling point at 1 bar [°C]	25–215	65
Density (STP) [kg/m^3]	740	790

Table 1.
Physical attributes of methanol and gasoline [39].

Property	Gasoline	Methanol
Chemical formula	Various	CH3OH
Adiabatic Flame Temperature [K]	2275	2143
Autoignition temperature [K]	465–743	738
Stoichiometric AFR [kg/kg]	14.7	6.5
Hydrogen Content by mass [%]	14	12.58
Volumetric Energy Content [MJ/m^3]	31,746	15,871
Higher heating value [MJ/kg]	48	22.88
Lower heating value [MJ/kg]	42.9	20.09
Carbon Content by mass [%]	86	37.48
Oxygen Content by mass [%]	0	49.93
Molecular weight [kg/kmol]	107	32.04

Table 2.
Chemical aspects of methanol and gasoline [39].

concerns. Methanol, itself, has excellent properties to be used as an optimum alternative fuel but without properly discussing the above-mentioned issues, its uses will be limited in the automotive sector and, hence, cannot reach its complete potential in this particular sector. The details of these issues and their predicted solutions have been mentioned in the later sections.

3.1 Cold start issue

Cold start still exists as one of the core challenges encountered while using methanol in SI engines. The major reasons due to which this issue arises in SI engines are the physio-chemical properties of methanol fuel, which differ greatly from the conventional gasoline. During the cold start, the components of the SI engine are cooler, thereby demanding an efficient vaporization process in order to propagate complete combustion of the fuel [40]. However, the high heat of vaporization of methanol presents a formidable obstacle, restraining the required atomization and vaporization for optimal fuel-air mixing in the early stages of the engine operation.

This particular issue of cold start gets worsened by lower volatility of methanol fuel, when compared to that of gasoline. This lessened volatility of methanol hinders the fuel's ability to readily evaporate at decreased temperatures, thus, obstructing the generation of a homogeneous and combustible fuel-air mixture. As a result, the spark ignition engines struggle to attain a complete and efficient combustion in the crucial ignition phase, therefore, resulting in the suboptimal engine performance, enhanced emissions and heightened fuel consumption [41]. Furthermore, the issue of cold start also gets worse by the presence of water in methanol fuel, whether as a contaminant or through condensation. This presence of water particles in methanol serves as an additional hindrance and thus slows down the fuel's vaporization process. This results in decreasing the engine's ability to obtain the required air-fuel mixture, therefore generating a negative impact on the overall efficiency and the effectiveness of the entire combustion process [42].

The cold start issue results in the consequences that reverberate throughout the engine processes. The incomplete combustion due to the cold starts leads to decreased power output, compromised efficiency of fuel and increased emissions, which act as a counter to the objective of achieving sustainable transportation [43]. The SI engines, as a result, may experience rough idling, misfires or hesitation till reaching the desired operating temperature, thereby, leading to discomfort for the driver as well as significant wear on the engine components.

In order to address the challenge of cold start for methanol fuel in SI engines, there is a need for effective and innovative solutions. One of the solutions that the researchers have suggested in recent years is engine modification for employing methanol fuel in SI engines. Various types of engine modifications have been identified to optimize the vaporization process of the fuel. The fuel vaporization and atomization can be enhanced by using high-pressure fuel injection systems. These systems will ensure the improved air-fuel mixing during the cold starts [44]. Moreover, electrically heated injectors can be used as an additional heat source, aiding the vaporization process and thus resulting in smoother ignition of engine in cold starts. Another modification is to utilize the preheated intake air for the combustion process which will elevate the fuel's temperature, therefore, strengthening the fuel's ability to vaporize efficiently and allowing the combustion process to be more effective during the cold starts.

Furthermore, the utilization of fuel additives can also serve as a potential remedy for this particular problem. The ignition enhancers, such as small amounts of gasoline and ether, can aid in starting the combustion process in cold starts, thereby facilitating easier ignition of the SI engines. The temperature of the methanol fuel can also be increased by the application of fuel heaters, engine coolant-heated systems as well as electrically powered heated systems, thus, enhancing its vaporization and improving the combustion efficiency from the very outset [45].

3.2 Enhanced NOx emissions

The implication of methanol as a sustainable alternative fuel for SI engines has gathered considerable attention because of its potential to significantly reduce the GHG emissions and reliance on conventional fossil fuels. However, a major concern linked with methanol's usage for SI engines lies in the enhanced nitrogen oxide (NOx) emissions.

There are several reasons for the heightened NOx emissions while utilizing methanol as a fuel in SI engines, out of which the primary reason is the contrasting physio-chemical properties of methanol when compared to conventional gasoline. For complete and efficient combustion process in SI engines, methanol requires a longer duration than gasoline because of its high latent heat of vaporization [46]. As a result, the extended combustion process leads to enhanced residence time of nitrogen molecules in the combustion chamber, thus, facilitating the generation of nitrogen oxides. Moreover, characterized by the excess air in the air-fuel mixture, the combustion of methanol exhibits a lean-burn tendency. While this condition of lean combustions illustrates the potential to increase fuel efficiency and reduce CO_2 emissions, it consequently promotes the formation of NOx emissions. Under these lean conditions, the availability of excess oxygen favors the generation of thermal NOx, where the molecules of nitrogen and oxygen react at increased temperatures to produce complex NOx compounds [47].

Apart from the intrinsic nature of the methanol, the combustion process also plays an important role when it comes to the NOx emissions. The formation of nitrogen oxides is exacerbated due to the high flame temperature associated with methanol combustion in spark ignition engines. The pathway for the formation of thermal NOx gets intensified due to the elevated temperatures as this enhanced temperature acts as a catalyst for the reaction between nitrogen and oxygen molecules. Consequently, the combustion of methanol in SI engines tends to produce substantially higher NOx emissions as compared to the neat gasoline [48].

To address this issue of enhanced NOx emissions when using methanol in SI engines, several mitigation strategies can be adopted. One of the strategies involves the application of catalytic converters equipped with selective catalytic reduction (SCR) systems [49]. These specific types of systems use a catalyst, mostly a metal such as platinum or palladium, for the conversion of NOx compounds into water and nitrogen *via* reduction reaction [50]. A promised reduction in NOx emissions has been shown by this technology and thus can be adapted for utilizing methanol in SI engines.

Another potential approach involves the optimization of combustion process for minimizing the formation of NOx. Various advanced management techniques such as timing adjustment of fuel injection and exhaust gas recirculation (EGR) can be used for controlling the temperature within the combustion chamber, thereby reducing the generation of NOx compounds [51].

3.3 100% pure methanol utilization

The utilization of 100% methanol in spark ignition engines presents one of the core challenges that emerges from a multitude of technical and logistical considerations. Methanol, a colorless volatile alcohol, has physio-chemical properties that differ significantly when compared to that of gasoline. The core challenge lies in reconciling the unique characteristics of methanol with the intricate requirements of SI engines, demanding innovative engineering and adaptation.

One of the primary hurdles in using 100% pure methanol in SI engines is its dissimilar combustion behavior compared to the neat gasoline. Methanol, as a fuel, possesses high octane rating, which is advantageous when it comes to knocking and resistance to pre-ignition. However, the downside of methanol is its slow flame propagation speed which generates the need for modifications for engine's injection and combustion systems [52]. With methanol, ensuring efficient combustion and effective power delivery requires precise calibration as well as fine tuning because the modified combustion attributes have significant impact on the several factors including, air-fuel ratio, ignition timing and injection strategy.

Furthermore, the low energy density of methanol as compared to gasoline has a direct negative impact on the overall efficiency of the engine. In order to maintain the power output, there is a need to increase the fuel's flow rate during ignition, which consequently results in enhanced fuel consumption. Thus, in order to provide the unaffected engine performance, large quantities of methanol are required which pose a challenge in terms of fuel storage and transport. As a result, the automotive engineers and engine manufacturers must tackle the challenge of fuel efficiency optimization while maintaining sufficient range and refueling infrastructure [53].

One of the many challenges that occurs in utilizing 100% pure methanol as a sustainable alternative fuel is its corrosive nature of properties. This corrosive nature of methanol damages certain metals, rubbers and polymers that are commonly used in several components of engines. This necessitates comprehensive material compatibility studies and careful selection of materials for manufacturing engine components [54]. Due to this particular nature of methanol, it is imperative for engine manufacturers to develop corrosion alloys as well as compatible seals and gaskets for mitigation of potential damage, thereby ensuring the durability of engine [55].

Moreover, the extensive usage of 100% pure methanol as a fuel, alternative to conventional gasoline, requires a comprehensive infrastructure overhaul. Due to the difference in distinctive handling, storage and transportation requirements of methanol, the currently employed gasoline distribution network is ill-suited to accommodate pure methanol [56]. The construction of such distribution network for the purpose of utilizing 100% pure methanol would require significant investments and time, thereby hindering its adoption on commercial scale [57]. Furthermore, in order to ensure the availability of methanol on refueling stations for consumers poses a substantial logistical challenge, demanding extensive cooperation and collaboration between fuel providers, vehicle manufacturers and regulatory authorities.

3.4 Transportations concerns

Various transportation issues also need to be addressed before the usage of methanol in SI engines on commercial scale. Methanol, possessing the potential of a renewable and clean-burning fuel, provides significant benefits when it comes to reduced emissions and reliance on conventional fossil fuels. However, there are several challenges associated with the distribution and transportation of methanol, which present quite a hindrance in its utilization in SI engines. The later sections will present the details of these challenges as well as the predicted solutions for overcoming them.

One of the major concerns lies in the infrastructure that is required for the transportation and storage of methanol. Methanol has contrasting properties when compared to the conventional fuels like gasoline and diesel, including its enhanced flammability and corrosiveness. Thus, the implementation of special materials and safety measures in transportation infrastructure seems imperative, including the pipeline tanker and other storage facilities [58]. This will not only ensure the safe handling and transport of methanol but also make the transportation faster than usual. However, such an advancement in infrastructure requires substantial capitals and comprehensive planning in order to support the ubiquitous adoption of methanol as an alternative to gasoline.

Another important transportation challenge involves the accessibility and availability of refueling stations of methanol. As of now, there is a limited access of consumers to methanol in terms of fuel stations as compared to that of neat gasoline. This creates the necessity of infrastructure development in order to facilitate the users of methanol-fueled applications, i.e., SI engines. Currently, they may be finding it very difficult to reach the methanol refueling stations, especially in remote or less populated regions [59]. Thus, the solution in this case will be expansion of methanol refueling stations network which will promote the widespread utilization of methanol as a renewable alternative fuel in SI engines.

In addition to above-mentioned challenges, methanol's transportation as well as its distribution on large scale also involves serious logistical considerations [60]. The production of methanol is mainly from sources such as natural gas or biomass and due to this particular reason, the production facilities are typically found in specific regions to ensure the availability of raw materials for methanol [61]. Transporting the generated methanol from these production facilities to different regions needs efficient as well as effective logistical planning, including the modes of transportation, routes and last but not least, storage facilities. For this particular challenge, the solution involves the development of robust supply chains and optimization of ongoing logistics. This will be very crucial to ensure the reliability and cost-effectiveness of methanol distribution to the end-users.

Moreover, methanol's energy density is less when compared to conventional gasoline. This means that more quantity of methanol is required to obtain the same energy content as that of gasoline, consequently, necessitating the need of large containers or tanks and specialized equipment for the accommodation of larger volumes of methanol [62]. To address this specific issue, there is a need of innovative solutions such as increasing the methanol's energy content *via* blending with some other conventional but high energy fuels or by exploring sophisticated means of storage and technological advancements related to delivery of methanol. When transporting such kind of corrosive and highly flammable, one significant concern is the safety factor involved. As it is mentioned previously that methanol is less flammable than gasoline, its higher flammability as compared to diesel originates certain safety challenges. Proper safety protocols as well as regulations are needed to be established and abided throughout the process of transportation, including, handling, loading, unloading and storage of methanol. The education and training of personnel involved in this entire procedure of transportation and distribution is necessary for ensuring safe practices as well as minimizing the risk of tragic accidents.

3.5 Lubricant oil deterioration

In the world of automotive engineering, the selection of fuel for a vehicle or any transport plays a very keen role in delivering optimized engine performance as well as engine longevity. As the entire world is shifting from conventional fossil fuel toward alternative fuels, such as methanol, it becomes imperative to understand the potential impacts of such fuels on lubricant oil degradation. The later sections have explained many parameters such as kinematic viscosity, total acid number, specific gravity and flash point of lubricant oil in order to determine the feasibility of methanol for engine's durability.

3.5.1 Kinematic viscosity

Kinematic viscosity is an important parameter which measures the resistance of lubricant to flow and thus plays a very crucial role in evaluating its effectiveness. Methanol demonstrates lower viscosity as compared to gasoline when it comes to the utilization of methanol as an alternative to gasoline in SI engines. This lessened viscosity is due to the unique molecular structure and contrasting physio-chemical properties of methanol [63, 64]. Methanol, as compared to gasoline, is a smaller molecule and, hence, possesses weaker intermolecular forces as well as lower molecular weight. The lower viscosity of methanol can be attributed to these previously mentioned properties. This decreased viscosity has a drastic impact on the lubricant oil, being used in SI engines, thereby, resulting in compromising the lubricating film thickness that is formed between the engine's moving parts. This reduced thickness of lubricating film eventually results in reduced lubrication, enhanced friction and heightened wear and tear between the moving components [64]. The consequences of this enhanced wear and increased friction can prove detrimental for the engine's performance and its durability. For addressing the challenges resulted due to reduced kinematic viscosity, specialized lubricants must be employed that are formulated to meet the specific requirements of utilizing methanol as a fuel in spark ignition engines. These specific properties will include enhanced film-formation and involvement of anti-wear additives for compensating the lower viscosity.

3.5.2 Total acid number

The total acid number (TAN) is an important property of lubricant oil and thus plays a key role in identifying the acid content present within the oil. When methanol is burnt in spark ignition engines, its combustion can produce several acidic compounds. These acidic by-products of methanol's combustion can infiltrate the lubricant oil through various channels and pipelines such as blow-by gases. Consequently, the TAN of the oil will increase, thereby indicating an increase in acidification. This availability of acidic compounds within the lubricant oil can lead to detrimental effects on components of engine. One of the major concerns is the enhanced corrosion, which eventually promotes the degradation of critical engine parts and lessened engine lifespan [65]. The corrosive nature of these acidic compounds can erode surfaces, impair proper functioning and sub-optimal engine performance. In order to reduce the impact of acidification, continuous monitoring of TAN and implementation of proper maintenance schedules are imperative [66]. This includes timely oil changes and the utilization of high-quality lubricants with suitable additives to neutralize the effect of acidic compounds, thus, maintaining a favorable environment for engine operation.

3.5.3 Specific gravity

Specific gravity refers to the density of a substance compared to the density of water. The specific gravity of methanol is lower when compared to traditional gasoline. This particular characteristic of methanol can have detrimental impacts on the performance of SI engines during methanol usage. The major consequence of this lower specific gravity is the reduction in oil film thickness which results in increased metal-to-metal contact inside the engine [67]. This enhances contact between the components of engine results in heightened friction and wear. This increased friction and wear can contribute to premature component deterioration, reduced engine efficiency and a shorter lifespan for critical engine parts. The lessened oil film thickness becomes a hindrance in lubrication effectiveness, allowing for direct metal contact and heightened friction between the moving parts [68]. These deleterious impacts of thinned oil film mark the importance of usage of special lubricants which are formulated in such a way that they can contribute for the lower specific gravity of methanol during its utilization in SI engines as a fuel. These measures will help maintain optimum lubrication between the components, reduce friction between them and thus mitigate the significant negative effects on the engine performance and lifespan.

3.5.4 Flash point

The flash point is an important property of lubricant oil, which indicates the minimum temperature at which sufficient vapor gets released from it, in order to ignite when exposed to an open flame or spark. When methanol is used in the SI engines instead of neat gasoline, the lubricant oil after designated hours of running exhibits lower flash point as compared to the traditional gasoline. Thus, the utilization of methanol enhances the potential for heightened operating temperature, thereby resulting in increased incidents of flash points that can occur within the lubricant oil [69]. This reduction in flash points can create significant impact on the chemical stability of oil, leading to unfortunate consequences [70]. One of its major consequences is the enhanced formation of deposits and sludge. The occurrence of flash points at higher frequencies can result in lubricant oil degradation, thus, formulating the by-products that promote the production of sludge and deposits. This accumulation of sludge and other deposits can generate overall negative impact on the engine's performance and durability by hindering the circulation of oil, inhibiting proper lubrication and lessening the effectiveness of several crucial components of engine [71]. To lessen the effects of frequent flash points, the use of special lubricant oils, which are formed to withstand increased temperatures and maintain their chemical stability, is imperative. These special oils much possess better resistance to thermal degradation and must be equipped with increased detergency and dispersant properties for the purpose of minimizing the formulation as well accumulation of sludge and deposits in lubricant oil.

4. Artificial neural network-based prediction

Techniques like artificial neural network (ANN) have risen as powerful tools for the prediction and analysis of complex patterns in different fields [72]. This technique can be employed to predict the utilization of gasoline-methanol blends in SI engines. Methanol can be described as a renewable and environmentally friendly fuel, which has garnered quite an attention as a wonderful alternative to conventional gasoline because of its reduced carbon footprint. However, for the purpose of its successful integration into the currently used SI engines, the accurate predictions of various engine characteristics including performance, emissions and combustion seem imperative. This is the space which can be filled with the application of ANN in a wonderful manner.

ANNs can be defined as computational models which are inspired and generated by the human brain's neural network structure [73]. They comprise of interconnected neurons (artificial) that can process and, hence, transmit information. The historical data acts as a raw material in training an ANN and it can learn to identify several patterns and relationships within the data, allowing it to predict the future data. In this particular context of utilizing gasoline-methanol blends in SI engines, based on various factors, ANNs can be trained to predict the behavior of various engine characteristics. In the beginning, ANN can be provided with the historical data about the chemical composition of gasoline-methanol blends which includes the percentage of methanol in the entire blend [74]. Other important parameters, including, ignition timing, engine speed, air-fuel ratio and engine load can also serve the purpose of input parameters. With the help of historical data that can provide information on the above-mentioned parameters, ANNs can learn to determine patterns and correlations. This enables the ANNs to create multiple predictions about the

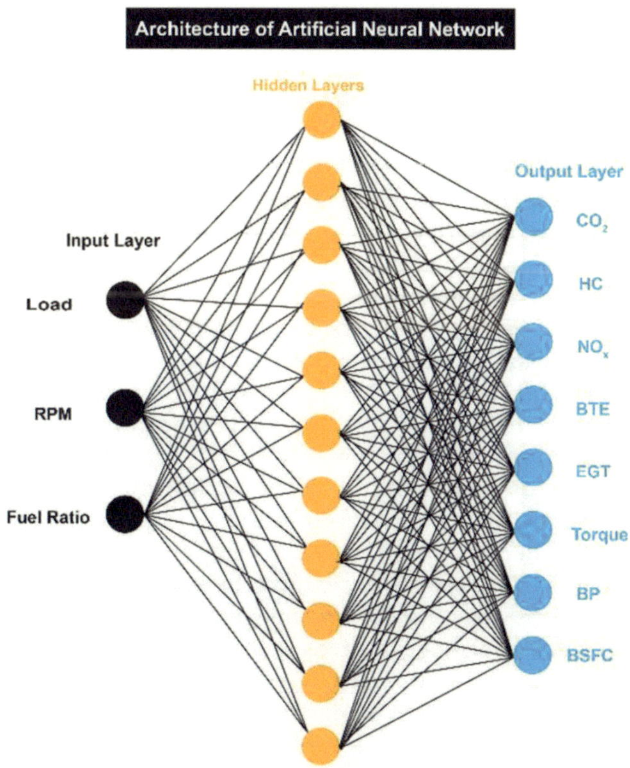

Figure 6.
Development of ANN model for gasoline-methanol blends.

engine's performance and emissions when provided with the updated values of input parameters [75]. **Figure 6** demonstrates the development of ANN model for various fuel blends. The input parameters include engine load, engine speed and fuel blends, of which the effects have been observed on output parameters such as performance and emission characteristics. The ability to capture nonlinear relationships between input and output variables can be described as one of the key advantages of using ANN techniques. It is often difficult for traditional mathematical models to identify the complex interaction within the engine system. Rather than this, ANNs excel at determining complicated relationships between multiple input and output variables, and hence can precisely provide the predictions about the impact of various gasoline-methanol blends composition on the engine performance, emissions and combustion parameters [76, 77].

By using this special technique, named ANN, researchers and engineers can discover a vast range of scenarios, without any particular need for comprehensive practical experimentation. They can easily analyze the effect of various ratios of gasoline-methanol blends on critical output parameters including the fuel efficiency, power output, generated torque as well as emission of several pollutants such as carbon monoxide (CO), oxides of nitrogen (NOx) and unburned hydrocarbons (HC). This predictive capability of this technique develops a model for efficient optimization of engine performance while ensuring compliance with emissions regulations [78]. Moreover, ANN can assist real-time forecasts, thereby, making them valuable in onboard control systems of engines. With the help of continuous surveillance on input variables and utilizing the modified ANN model, the operating conditions of engine can be adjusted by the system dynamically for optimize the performance as well as emissions, based on the desired composition of gasoline-methanol blend [79].

4.1 Comparison with other prediction techniques

In addition to artificial neural network (ANN), several other deep learning architectures such as convolutional artificial network (CNN) and recurrent artificial network (RNN) can also be used for the prediction analysis of gasoline-methanol blends. ANN can be defined as a foundational model that comprises of interconnected nodes or neurons, arranged in the form of layers. Each neuron has the capability to process input data and *via* iterative learning as well as past records, ANN can optimize its weights to capture intricate patterns in the provided data. However, despite being a wonderful technique, ANN lacks spatial awareness, which makes it less suitable for tasks including, grid-like data such as images.

Unlike ANN, convolutional neural network (CNN) is particularly designed for image processing tasks. Its basic function involves employment of multiple convolutional layers to identify spatial patterns and features in the provided data. These layers use versatile filters in order to scan the input and thus acquire the local information, thereby allowing the CNNs to effectively extract the meaningful information about the features and recognize objects inside the images. In comparison with ANN, CNN's architecture enhances the weight sharing, as well as lessens the model's memory demands and increases its capability to handle more intricate visual data. In comparison, recurrent artificial network (RNN) can be used effectively for sequential data, including the time series data. It comprises of a recurrent collection that generates feedback loops, thus enabling the RNN to capture information from previous time steps. The prediction analysis based on the sequential dependencies can be easily handled by this temporary memory of RNN. The ability of RNN to model dynamic patterns makes it an ideal candidate for forecasting tasks including temporal data such as prediction of gasoline-methanol blends over time.

To conclude, it can be said that a versatile model like ANN is best suited for complicated relationships. CNNs excel in image-processing tasks, whereas RNN can be used proficiently for capturing temporal dependencies. Combining these different models can allow us to have a comprehensive approach to forecasting analysis of gasoline-methanol blends, leveraging their individual strengths for addressing various aspects of the data.

5. Optimization

Determining the optimized conditions for the utilization of gasoline-methanol blends in SI engines is a complex task which requires a comprehensive and systematic approach. One method that is quite effective and has been widely used for this purpose is response surface methodology (RSM). RSM can be defined as a statistical and mathematical tool that can be employed for modeling and optimization of the relationships between multiple input variables and a response or output variable [80]. One of the basic steps in using RSM is the selection of input variables that creates an impact on the performance characteristics of methanol-gasoline blends [81]. The examples of such factors can include blend ratio of gasoline to methanol, ignition timing, engine speed, load, compression ratio, air-fuel ratio and other several associated parameters. The factors that are chosen in the first stage should cover a wide range of values for the purpose of capturing the full spectrum of engine processes. In the next step, a set of experiments is performed for collecting data on the chosen input variables and the corresponding output variable(s) [82]. The response variables can be any of the engine's performance characteristics including torque, brake power, brake-specific fuel consumption, brake thermal efficiency as well as the corresponding emissions of the engine. The design of experiment must be planned carefully so that it can ensure enough number of data points that can easily cover the design space as well as give reliable information for modeling. After the collection of the data attained *via* experimentation, RSM will generate a mathematical model using the regression analysis for describing the relationship between input and output variables [83].

Now, with the availability of mathematical model, the next step is the optimization of conditions for using gasoline-methanol blends in SI engines. RSM employs optimization algorithms, including the desirability function approach, to find the combination of input variables that can either maximize the required response variables or minimize the undesirable ones. The values of desirability function range from 1, representing the best, to 0, signifying the worst [84]. In RSM, the optimization process includes adjustment of values of the input variables iteratively, that is based on the generated mathematical model and the required optimization criteria. Repeated evaluations of the model are done, and it gets updated until the optimization of desired response [85]. This procedure of iterations helps researchers to systematically explore the design space and evaluate the optimal conditions for using gasoline-methanol blends in SI engines. **Figure** 7 shows an example of optimization process for fuel blends. The desirability chart of optimization has been generated by RSM for showing the contribution of each response variable in developing the optimized model for the engine.

There are several significant benefits that researchers will get by using RSM for optimizing the conditions of gasoline-methanol blends. First of all, with a relatively

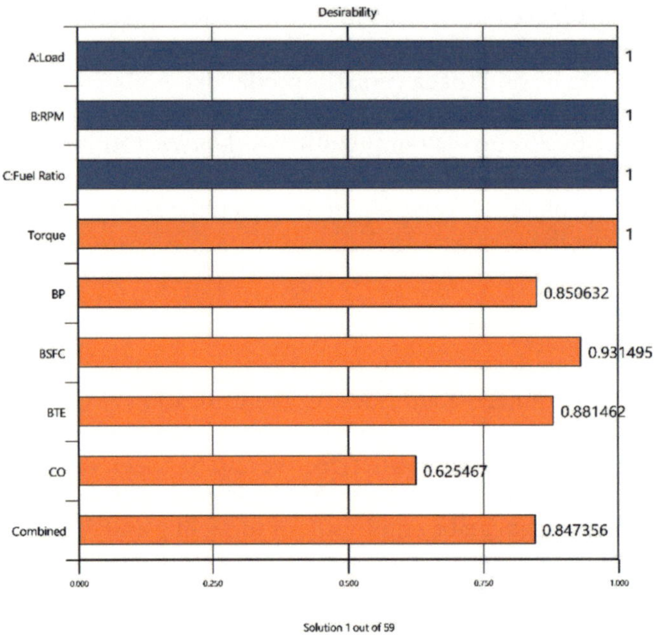

Figure 7.
Desirability chart.

limited number of experimentations, RSM can provide the efficient exploration of a large parameter space. This saves time, resources and cost compared to conducting a large number of exhaustive experiments. Secondly, a systematic as well as a structured approach can be generated with the help of RSM, which allows the understanding of complex relationships between input and response variables (engine performance and emissions) [86].

5.1 Comparison with other optimization techniques

Besides response surface methodology (RSM), several other distinct techniques including the Taguchi Method and Genetic Algorithm can also be used in an effective manner for the purpose of optimization of gasoline-methanol blends. RSM can be defined as a statistical approach that strives for modeling and optimizing the relation between multiple input variables (engine speed, load, blend ratio) and the response variables (performance and emission characteristics). RSM helps in capturing the interaction between different variables and thus aids in locating the optimal operating conditions for the superior gasoline-methanol blends. The Taguchi Method, developed by Genichi Taguchi, emphasizes on reliable optimization by lessening the effects of noise and variations in manufacturing or experimental processes. This particular method uses an orthogonal array for efficiently conducting the experiments with a limited set of trials, thereby, making it less time-consuming as well as cost-effective. The major purpose of Taguchi Method is to find the optimal variable settings that lead to minimizing the sensitivity of responses to external variations, thus, enhancing the reliability and sensitivity of the gasoline-methanol blends. On the other hand, genetic algorithm (GA) can be defined as an evolutionary optimization technique which is inspired by the laws of natural selection and genetics. It initiates with a set of potential solutions, termed as chromosomes, which represent different combinations of input variables as defined above. With the help of successive generations, GA implements selection, cross-over and mutation techniques to evolve innovative solutions over time. GA is mostly useful when dealing with the intricate, nonlinear and multi-dimension optimization problems. This makes it a suitable candidate for deriving optimal gasoline-methanol blends in a vast parameter space.

To conclude, it can be said that RSM focuses majorly on statistical modeling for deriving the optimal settings of input variables. On the other hand, the Taguchi Method emphasizes on robust, stable and least sensitive optimization. In comparison, GA utilizes evolutionary principles in order to search for solutions in complex and multi-dimensional spaces, considering nonlinear relationships. With complete understanding about the strengths as well as limitations of above-mentioned methods, engineers and researchers can make better decisions about which approach is the most suitable one for gasoline-methanol optimization objectives.

6. Environment and sustainable development goals (SDGs)

The sustainable development goals (SDGs) emphasize the importance of maintaining balance between social involvement, economic growth and environment protection for ensuring the development of a sustainable era. These goals provide a roadmap for different governing bodies, institution and individuals to work cooperatively toward a most inclusive, just and sustainable world. Utilization of methanol in SI engines as a fuel, instead of neat gasoline, allows us to achieve some important sustainable development goals which include, good health and well-being, climate action, clean and affordable energy and responsible consumption and production.

First of all, an alternative fuel like methanol has the potential of substantially reducing GHG emission when compared to traditional gasoline. The combustion of methanol in SI engines will produce lower levels of CO_2 and other hazardous pollutants including the oxides of nitrogen (NOx) and particulate matter (PM). With the utilization of methanol as a fuel in SI engines, we can surely and actively contribute to sustainable development goal of climate action, by lessening the effect of climate change and reducing the carbon footprint in transportation sector. Secondly, the usage of methanol in SI engines can generate a positive impact on another sustainable development goal, i.e., good health and well-being. The usage of traditional gasoline in SI engines emits large number of hazardous pollutants which have deleterious effects on human health, thereby, contributing to the increase in respiratory diseases and other diseases. The utilization of methanol not only produces less harmful emissions but can also lead to improved air quality. This will eventually lead to lessened health risks linked with vehicular emissions. Thus, by using methanol as a fuel, a healthier and safer environment can be created for upcoming generations and communities.

Moreover, since methanol is a renewable, clean and cost-effective source, this aligns with the sustainable goal of clean and affordable energy. Various kinds of feedstocks, including renewable resources such as captured carbon dioxide and biomass, can be used for the production of methanol. This versatility in feedstocks allows for sustainable and clean production as well as helps in the integration into the existing energy platforms. Thus, by utilizing methanol in SI engines, we can make our energy resources more versatile and reduce reliance on conventional fossil fuels, thereby contributing to the cleaner and renewable energy future.

In addition to the above-mentioned SGDs, the adoption of methanol as an alternative to gasoline promotes the sustainable development goal of responsible consumption and production. Methanol can be easily produced *via* environmentally sensible methods and feedstocks, which lessen their impact on the ecosystems and thus support sustainable practices. Moreover, the efficiency of methanol's combustion properties and its compatibility with current engine infrastructure promote responsible consumption and its effective resource utilization. By using methanol as a fuel, the sustainable responsible consumption patterns can be encouraged, and this will lead to more environmentally friendly production practices.

7. Conclusions

Methanol, a promising alternative to neat gasoline, has been under observation for quite some time. Its physio-chemical properties have made it quite an attention-grabbing fuel for the research to address the issue of energy demands as well as environmental deterioration. However, there are certain core challenges that must be addressed before the utilization of methanol on commercial scale in the transportation sector.

Cold start issue stands as a prominent challenge when it comes to its utilization in SI engines as a fuel. Methanol's high heat of vaporization, lower volatility and the presence of water impede the fuel's atomization, vaporization and combustion during engine ignition phase. These particular attributes of methanol serve as an obstruction because they lead to reduced engine performance, intensified emissions and compromise fuel efficiency. However, with the application of fuel additives, the issue of cold start can be mitigated. Also, by increasing the fuel vaporization and enabling optimal air-fuel mixing, these innovative solutions can pave the way for an effective and sustainable utilization of the methanol in spark ignition engines.

The enhanced emissions of nitrogen oxides (NOx) also present a notable challenge in its usage as an alternative fuel in SI engines. The formation of NOx can be resulted due to peculiar combustion attributes of methanol, which includes elevated flame temperatures, extended combustion duration and lean-burn tendencies. However, by the application of several strategies like catalytic converters, it seems possible to address the issue of enhanced NOx emissions.

The utilization of 100% pure methanol in spark ignition engines seems quite impossible. However, the modification of engine's fuel delivery and ignition systems in order to accommodate the unique attributes of methanol, including its lower energy content and bizarre combustion characteristics can be favorable. Moreover, another solution is the application of fuel blending techniques for the preparation of ideal methanol-gasoline blends. This approach will not only enhance the combustion efficiency of methanol but also improve its energy content. In addition, modern engine technologies including direct injection and variable valve timing, can be utilized for the purpose of optimizing the performance of 100% pure methanol.

For commercial-scale utilization of methanol in spark ignition engines, the transportation sector needs to focus on infrastructure development, expansion of refueling station network, optimization of logistical section, ensuring safety and addressing of issues related to energy density. The collaboration among the industry stakeholders, policy makers and researchers will play an essential role in addressing the very issue. This will direct the full potential of methanol as a clean-burning, sustainable and efficient fuel for SI engines, thereby contributing to a cleaner and more sustainable transportation sector.

For the issue of lubricant oil deterioration, there is need of more research and development for formulating specialized lubricant oils which can significantly lessen their deterioration during the utilization of methanol in spark ignition engines. These specialized lubricant oils must possess enhanced film-formation properties, suitable additives to neutralize the effect of acidic compounds and better resistance to thermal degradation.

With the use of ANN techniques, a marvelous potential for predicting the utilization of gasoline-methanol blends in SI engines can be identified. ANNs can encompass intricate relationships, ensuring the accuracy of predictions about the several engine characteristics, based on the input variables. With the advancements in ANN techniques, it will surely play a very vital role in the development of sustainable and efficient engine systems for the upcoming generations. Moreover, RSM can be considered as a powerful tool in determining the optimal conditions for the utilization of methanol-gasoline blends in SI engines. This is achieved through systematic exploration of the design space and the utilization of mathematical models derived from the experimental data.

The utilization of methanol in SI engines provides an optimistic pathway for addressing the multiple sustainable development goals. By reducing the impact of climate change, ameliorating the air quality, making the energy resources more versatile and promoting the reliable consumption and maintaining the renewable production, methanol contributes to SGDs of climate action, good health and well-being, clean and affordable energy and responsible consumption and production. However, it is essential to make sure of sustainable production methods, effective distribution platforms and suitable waste management for maximizing the environmental and social perks of methanol usage. Thus, embracement of methanol as an alternative to gasoline in SI engines can promote a more sustainable and greener future.

Acknowledgements

The authors extend their appreciation to the Researchers Supporting Project number (RSPD2023R698), King Saud University, Riyadh, Saudi Arabia, for funding this work.

Author details

Muhammad Usman[1*], Muhammad Kashif Jamil[1], Ahsan Hanif[1], Muhammad Mujtaba Abbas[1], Mahir Es-Saheb[2] and Yasser Fouad[2]

1 Department of Mechanical Engineering, University of Engineering and Technology, Lahore, Pakistan

2 Department of Applied Mechanical Engineering, College of Applied Engineering, Muzahimiyah Branch, King Saud University, Riyadh, Saudi Arabia

*Address all correspondence to: muhammadusman@uet.edu.pk

References

[1] Singh S, Bansal P, Bhardwaj N. A study on nexus between renewable energy and fossil fuel consumption in commercial sectors of United States using wavelet coherence and quantile-on-quantile regression. Frontiers in Energy Research. 2022;**10**:848301. DOI: 10.3389/fenrg.2022.848301

[2] Mayer A. Fossil fuel dependence and energy insecurity. Energy, Sustainability and Society. 2022;**12**(1):27. DOI: 10.1186/s13705-022-00353-5

[3] Bozuwa J, Burke M, Cox S, Skandier CS. Chapter 8 - democratic governance of fossil fuel decline. In: Nadesan M, Pasqualetti MJ, Keahey J, editors. Energy Democracies for Sustainable Futures. Amsterdam, The Netherlands: Elsevier B.V, Academic Press; 2023. pp. 73-82. DOI: 10.1016/B978-0-12-822796-1.00008-5

[4] Yıldız İ. 1.12 fossil fuels. In: Dincer I, editor. Comprehensive Energy Systems. Oxford: Elsevier; 2018. pp. 521-567. DOI: 10.1016/B978-0-12-809597-3.00111-5

[5] Mukherjee A, Patel V, Shah MT, Munshi NS. Chapter 29- enzymatic and microbial biofuel cells: Current developments and future directions. In: Sahay S, editor. Handbook of Biofuels. Amsterdam, The Netherlands: Elsevier B.V, Woodhead Publishing; 2022. pp. 551-576. DOI: 10.1016/B978-0-12-822810-4.00029-4

[6] Speight JG. 2 - production, properties and environmental impact of hydrocarbon fuel conversion. In: Khan MR, editor. Woodhead Publishing Series in EnergyAdvances in Clean Hydrocarbon Fuel Processing. Amsterdam, The Netherlands: Elsevier B.V, Woodhead Publishing; 2011. pp. 54-82. DOI: 10.1533/9780857093783.1.54

[7] Hanif MA, Nadeem F, Tariq R, Rashid U. Chapter 1 - energy resources and utilization. In: Hanif MA, Nadeem F, Tariq R, Rashid U, editors. Renewable and Alternative Energy Resources. Amsterdam, The Netherlands: Elsevier B.V, Academic Press; 2022. pp. 1-30. DOI: 10.1016/B978-0-12-818150-8.00011-3

[8] Hou H, Lu W, Liu B, Hassanein Z, Mahmood H, Khalid S. Exploring the role of fossil fuels and renewable energy in determining environmental sustainability: Evidence from OECD countries. Sustainability. 2023;**15**(3):2048. DOI: 10.3390/su15032048

[9] Adekomaya O, Adama K, Okubanjo A. Bridging environmental impact of fossil fuel energy: The contributing role of alternative energy. Journal of Engineering Studies and Research. 2018;**23**:22-27. DOI: 10.29081/jesr.v23i2.267

[10] Chmielewski A. Environmental Effects of Fossil Fuel Combustion. Oxford, UK: UNESCO - Encyclopedia Life Support Systems (UNESCO-EOLSS); 2005

[11] The International Energy Agency IEA. Key World Energy. The International Energy Agency IEA. France: International Energy Agency; 2015. p. 82

[12] Martins F, Felgueiras C, Smitkova M, Caetano N. Analysis of fossil fuel energy consumption and environmental impacts in European countries. Energies (Basel). 2019;**12**:964. DOI: 10.3390/en12060964

[13] Holechek JL, Geli HME, Sawalhah MN, Valdez R. A global assessment: Can renewable energy replace fossil fuels by 2050? Sustainability. 2022;**14**(8):4792. DOI: 10.3390/su14084792

[14] Darkwah WK et al. Greenhouse effect: Greenhouse gases and their impact on global warming. Journal of Scientific Research and Reports. 2018;**17**:1-9. DOI: 10.9734/JSRR/2017/39630

[15] Wuebbles D, Jain A. Concerns about climate change and the role of fossil fuel use. Fuel Processing Technology. 2001;**71**:99-119. DOI: 10.1016/S0378-3820(01)00139-4

[16] Jeffry L, Ong M, Saifuddin NM, Rahman MM, Mubashir M, Show P-L. Greenhouse gases utilization: A review. Fuel. 2021;**301**:121017. DOI: 10.1016/j.fuel.2021.121017

[17] Valavanidis A. Global Warming and Climate Change. Fossil Fuels and Anthropogenic Activities Have Warmed the Earth's Atmosphere, Oceans, and Land. Hoboken, New Jersey, United States: Environmental Toxicology and Chemistry (ET&C) aligned with John Wiley & Sons, Inc; 2022

[18] Audi M, Ali A, Kassem M. Greenhouse gases: A review of losses and benefits. International Journal of Energy Economics and Policy. 2020;**10**:403-418. DOI: 10.32479/ijeep.8416

[19] Gurram A. Studies on spark ignition engine-a review. International Journal of Thermal Technologies. 2016;**6**:266-271

[20] Han S. Theoretical analysis of a spark ignition engine by the thermodynamic engine model. Journal of Energy Engineering. 2015;**24**:55-60. DOI: 10.5855/ENERGY.2015.24.3.055

[21] Shadloo MS, Poultangari R, Jamalabad MY. A new and efficient mechanism for spark ignition engines. Energy Conversion and Management. 2015;**96**:418-429. DOI: 10.1016/j.enconman.2015.03.017

[22] Mikhaylov A, Moiseev N, Aleshin K, Burkhardt T. Global climate change and greenhouse effect. Entrepreneurship and Sustainability Issues. 2020;7:2897-2913. DOI: 10.9770/jesi.2020.7.4(21)

[23] Yoro K, Daramola MO. CO_2 emission sources, greenhouse gases, and the global warming effect. In: Advances in Carbon Capture. Amsterdam, Netherland: Elsevier, Woodhead Publishing; 2020. pp. 1-28. DOI: 10.1016/B978-0-12-819657-1.00001-32

[24] Shahzad U. Global Warming: Causes, Effects and Solutions. Australia: Durreesamin Journal; 2015

[25] Methanol as a renewable fuel-a knowledge synthesis Report from an f3 project. 2017. [Online]. Available from: www.f3centre.se

[26] Dalena F, Senatore A, Marino A, Gordano A, Basile M, Basile A. Methanol production and applications: An overview. In: Methanol. Amsterdam, Netherland: Elsevier; 2018. pp. 3-28. DOI: 10.1016/B978-0-444-63903-5.00001-7

[27] Su L-W, Li X-R, Sun Z-Y. The consumption, production and transportation of methanol in China: A review. Energy Policy. 2013;**63**:130-138. DOI: 10.1016/j.enpol.2013.08.031

[28] Chen Y-H, Wong D, Chen Y-C, Chang C-M, Chang H. Design and performance comparison of methanol production processes with carbon dioxide utilization. Energies (Basel). 2019;**12**:4322. DOI: 10.3390/en12224322

[29] Marjani A, Rezakazemi M, Shirazian S. Simulation of methanol production process and determination of optimum conditions. Oriental Journal of Chemistry. 2012;**28**:145-151. DOI: 10.13005/ojc/280121

[30] Ramya V, Kodeeswararamanathan M. Methanol based fuel cell for power generation and its production from renewable sources. Transactions of the SAEST (Society for Advancement of Electrochemical Science and Technology). 2001;**36**:105-107

[31] Pearson RJ et al. Energy storage via carbon-neutral fuels made from CO_2, water, and renewable energy. Proceedings of the IEEE. 2012;**100**:440-460. DOI: 10.1109/JPROC.2011.2168369

[32] Bazaluk O, Havrysh V, Nitsenko V, Balezentis T, Streimikiene D, Tarkhanova E. Assessment of green methanol production potential and related economic and environmental benefits: The case of China. Energies (Basel). 2020;**13**:3113. DOI: 10.3390/en13123113

[33] Chen Z, Shen Q, Sun N, Wei Z. Life cycle assessment of typical methanol production routes: The environmental impacts analysis and power optimization. Journal of Cleaner Production. 2019;**220**:408-416. DOI: 10.1016/j.jclepro.2019.02.101

[34] K. Hashimoto and H. Chishima, E-methanol production using renewable hydrogen and its application to F1 fuel production. In: Conference on The 102nd CSJ Annual Meeting (2022). [Online]. Mar 2022

[35] Pellegrini L, Soave G, Gamba S, Langé S. Economic analysis of a combined energy-methanol production plant. Applied Energy. 2011;**88**:4891-4897. DOI: 10.1016/j.apenergy.2011.06.028

[36] Kralj A. Selecting different raw materials for methanol production using an MINLP model. Journal of Natural Gas Science and Engineering. 2014;**19**:221-227. DOI: 10.1016/j.jngse.2014.05.008

[37] Shirazi L, Aghaie E. Analysis of the methanol industry and its application from eco nomic standpoint. Petroleum and Coal. 2017;**59**:257-264

[38] Deka TJ, Osman A, Baruah D, Rooney D. Methanol fuel production, utilization, and techno-economy: A review. Environmental Chemistry Letters. 2022;**20**:3525-3554. DOI: 10.1007/s10311-022-01485-y

[39] Vancoillie J. Modeling the Combustion of Light Alcohols in Spark-Ignition Engines. Ghent, Belgium: IEEE, Department of Flow, heat and combustion mechanics; 2013

[40] Gong C, Wang S, Deng B, Su Y, Liu X. Effects of ignition and injection timing on cold start behavior of methanol engine. Energy. 2008;**42**:1367-1371

[41] Montaner Ríos G, Schirmer J, Becker F, Bleeck S, Gentner C, Kallo J. Automotive cold start of a PEMFC system by using a methanol solution as antifreeze. ECS Meeting Abstracts. 2020;**MA2020-02**:2184. DOI: 10.1149/MA2020-02342184mtgabs

[42] Li Z et al. Critical firing and misfiring boundary in a spark ignition methanol engine during cold start based on single cycle fuel injection. Energy. 2015;**89**:236-243. DOI: 10.1016/j.energy.2015.07.071

[43] Gong C, Fuxing W, Si X, Liu F. Effects of injection timing of methanol and LPG proportion on cold start characteristics of SI methanol engine with LPG enriched port injection

under cycle-by-cycle control. Energy. 2019;**144**:54-60

[44] Zhang B, Ji C, Wang S, Xiao Y. Investigation on the cold start characteristics of a hydrogen-enriched methanol engine. International Journal of Hydrogen Energy. 2014;**39**:14466-14471. DOI: 10.1016/j.ijhydene.2014.04.012

[45] Gong C, Fuxing W, Si X, Liu F. Effects of injection timing of methanol and LPG proportion on cold start characteristics of SI methanol engine with LPG enriched port injection under cycle-by-cycle control. Energy. 2017;**144**:54-60. DOI: 10.1016/j.energy.2017.12.013

[46] Iliev S. A comparison of ethanol and methanol blending with gasoline using a 1-D engine model. Procedia Engineering. 2015;**100**:1013-1022. DOI: 10.1016/j.proeng.2015.01.461

[47] Sileghem L, Huylebroeck T, Bulcke A, Vancoillie J, Verhelst S. Performance and emissions of a SI engine using methanol-water blends. SAE Technical Papers. 2013;**2**:126349. DOI: 10.4271/2013-01-1319

[48] Rifal M, Sinaga N. Impact of methanol-gasoline fuel blend on the fuel consumption and exhaust emission of a SI engine. AIP Conference Proceedings. 2016;**1725**:20070. DOI: 10.1063/1.4945524

[49] Babazadeh Shayan S, Seyedpour S, Ommi F, Moosavy SH, Alizadeh M. Impact of methanol-gasoline fuel blends on the performance and exhaust emissions of a SI engine. International Journal of Automotive Engineering. 2011;**1**:219-227

[50] Krishnaiah R, Porpatham E, Alexander J. Effects of methanol substitution on performance and emission in a LPG-fueled SI engine. In: Gupta A, Mongia H, Chandna P, Sachdeva G, editors. Advances in IC Engines and Combustion Technology. Singapore: Springer; 2021. pp. 193-205. DOI: 10.1007/978-981-15-5996-9_16

[51] Zhang Z et al. Effects of methanol application on carbon emissions and pollutant emissions using a passenger vehicle. PRO. 2022;**10**:525. DOI: 10.3390/pr10030525

[52] Agarwal A, Valera H, Pexa M, Čedík J. Introduction of methanol: A sustainable transport fuel for SI engines. In: Agarwal AK, Valera H, Pexa M, Čedík J, editors. Methanol. Energy, Environment, and Sustainability. Singapore: Springer; 2021. pp. 3-7. DOI: 10.1007/978-981-16-1224-4_1

[53] Çelik MB, Özdalyan B, Alkan F. The use of pure methanol as fuel at high compression ratio in a single cylinder gasoline engine. Fuel. 2011;**90**:1591-1598. DOI: 10.1016/j.fuel.2010.10.035

[54] Stelmasiak Z, Pietras D. Utilization of waste glycerin to fuelling of spark ignition engines. IOP Conference Series: Materials Science and Engineering. 2016;**148**:12087. DOI: 10.1088/1757-899X/148/1/012087

[55] Yellapragada D, Budda G. Methanol—A sustainable fuel for SI engine. In: Agarwal AK, Valera H, Pexa M, Čedík J, editors. Methanol. Energy, Environment, and Sustainability. Singapore: Springer; 2021. pp. 103-137. DOI: 10.1007/978-981-16-1224-4_5

[56] Bansal P, Meena R. Methanol as an alternative fuel in internal combustion engine: Scope, production, and limitations. In: Agarwal AK, Valera H, Pexa M, Čedík J, editors. Methanol. Energy, Environment, and Sustainability. Singapore: Springer; 2021. pp. 11-36. DOI: 10.1007/978-981-16-1224-4_2

[57] Verhelst S, Turner J, Sileghem L, Vancoillie J. Methanol as a fuel for internal combustion engines. Progress in Energy and Combustion Science. 2019;**70**:43-88. DOI: 10.1016/j.pecs.2018.10.001

[58] Azar CA, Lindgren K, Andersson B. Hydrogen or Methanol in the Transportation Sector?. United States: College of Information Sciences and Technolog; 2000

[59] López-Molina A, El-Halwagi D. An integrated approach to the design of centralized and decentralized biorefineries with environmental, safety, and economic objectives. PRO. 2020;**8**:1682. DOI: 10.3390/pr8121682

[60] Dehane A, Boumediene H, Merouani S, Hamdaoui O. The impact of methanol mass transport on its conversion for the production of hydrogen and oxygenated reactive species in sono-irradiated aqueous solution. Ultrasonics Sonochemistry. 2023;**95**:106380. DOI: 10.1016/j.ultsonch.2023.106380

[61] Kianfar E, Mazaheri H. Methanol to Gasoline: A Sustainable Transport Fuel. Hauppauge, New York, United States: Nova Science; 2020

[62] Müller K, Fabisch F, Arlt W. Energy transport and storage using methanol as a carrier. Green. 2014;**4**. DOI: 10.1515/green-2013-0028

[63] Kumar V, Agarwal A. Friction, wear, and lubrication studies of alcohol-fuelled engines. In: Kumar V, Agarwal AK, Jena A, Upadhyay RK, editors. Advances in Engine Tribology. Energy, Environment, and Sustainability. Singapore: Springer; 2022. pp. 9-29. DOI: 10.1007/978-981-16-8337-4_2

[64] Usman M et al. Experimental assessment of regenerated lube oil in spark-ignition engine for sustainable environment. Advances in Mechanical Engineering. 2020;**12**:3-8. DOI: 10.1177/1687814020940451

[65] Usman M et al. Comparative assessment of ethanol and methanol–ethanol blends with gasoline in SI engine for sustainable development. Sustainability. 2023;**15**:1-21. DOI: 10.3390/su15097601

[66] Usman M, Naveed A, Saqib S, Hussain J, Kashif M. Comparative assessment of lube oil, emission and performance of SI engine fueled with two different grades octane numbers. Journal of the Chinese Institute of Engineers. 2020;**43**:1-8. DOI: 10.1080/02533839.2020.1777205

[67] Jamil M et al. Analysis of the impact of propanol-gasoline blends on lubricant oil degradation and spark-ignition engine characteristics. Energies (Basel). 2022;**15**:5757. DOI: 10.3390/en15155757

[68] Ijaz Malik MA et al. Response surface methodology application on lubricant oil degradation, performance, and emissions in SI engine: A novel optimization of alcoholic fuel blends. Science Progress. 2023;**106**:26-33. DOI: 10.1177/00368504221148342

[69] Usman M, Saleem MW, Saqib S, Umer J, Naveed A, Hassan Z. SI engine performance, lubricant oil deterioration, and emission: A comparison of liquid and gaseous fuel. Advances in Mechanical Engineering. 2020;**12**:6-9. DOI: 10.1177/1687814020930451

[70] Ijaz Malik MA, Usman M, Hayat N, Hassan Zubair SW, Bashir R, Ahmed E. Experimental evaluation of methanol-gasoline fuel blend on performance, emissions and lubricant oil

deterioration in SI engine. Advances in Mechanical Engineering. 2021;**13**:13-14. DOI: 10.1177/16878140211025213

[71] Chowdary K, Tated M, Kotia A. Effect of Methanol and Ethanol on Lubrication Oil Degradation of CI Engine. Mumbai, India: Gujarat Research Society; 2019

[72] Dhande D, Choudhari CS, Gaikwad DP, Dahe K. Development of artificial neural network to predict performance of spark ignition engine fuelled with waste pomegranate ethanol blends. Information Processing in Agriculture. 2022. DOI: 10.1016/j.inpa.2022.05.001

[73] Kiani Deh Kiani M, Ghobadian B, Tavakoli T, Nikbakht AM, Najafi G. Application of artificial neural networks for the prediction of performance and exhaust emissions in SI engine using ethanol- gasoline blend. Energy. 2010;**35**:65-69. DOI: 10.1016/j.energy.2009.08.034

[74] Shukri MR, Rahman M, Devarajan R, Kadirgama K. Artificial neural network optimization modeling on engine performance of diesel engine using biodiesel fuel. International Journal of Automotive and Mechanical Engineering. 2015;**11**:2332-2347. DOI: 10.15282/ijame.11.2015.15.0196

[75] Bhatt A, Shrivastava N. Application of artificial neural network for internal combustion engines: A state of the art review. Archives of Computational Methods in Engineering. 2021;**29**:897-919. DOI: 10.1007/s11831-021-09596-5

[76] Ahmed E, Usman M, Anwar S, Ahmad HM, Nasir M, Ijaz Malik MA. Application of ANN to predict performance and emissions of SI engine using gasoline-methanol blends. Science Progress. 2021;**104**:15-21. DOI: 10.1177/00368504211002345

[77] Ye J. Artificial neural network modeling of methanol production from syngas. Petroleum Science and Technology. 2019;**37**:1-4. DOI: 10.1080/10916466.2018.1560321

[78] Banerjee R, Roy S. Development of an artificial neural network based virtual sensing platform for the simultaneous prediction of emission-performance-stability parameters of a diesel engine operating in dual fuel mode with port injected methanol. Energy Conversion and Management. 2019;**184**:488-509. DOI: 10.1016/j.enconman.2019.01.087

[79] Thangavelu SK, Ahmed A, Ani F. Application of ANN to predict S.I. Engine performance and emission characteristics Fuelled bioethanol. Applied Mechanics and Materials. 2014;**554**:454-458. DOI: 10.4028/www.scientific.net/AMM.554.454

[80] Yesilyurt M, Uslu S, Yaman H. Modeling of a port fuel injection spark-ignition engine with different compression ratios using methanol blends with the response surface methodology. Proceedings of the Institution of Mechanical Engineers, Part E: Journal of Process Mechanical Engineering. 2022;**237**(3):936-944. DOI: 10.1177/09544089221112373

[81] Ismail MY, Mamat R, Azmi W, Awad O, Ali O. Application of response surface methodology in optimization of performance and exhaust emissions of secondary butyl alcohol-gasoline blends in SI engine. Energy Conversion and Management. 2017;**133**:178-195. DOI: 10.1016/j.enconman.2016.12.001

[82] Taymaz I, Akgun F, Benli M. Application of response surface methodology to optimize and investigate the effects of operating conditions on the performance of DMFC. Energy. 2011;**36**:1155-1160. DOI: 10.1016/j.energy.2010.11.034

[83] Yellapragada D, Rao B, Vedula D, Anusha C. Improvement of biodiesel methanol blends performance in a variable compression ratio engine using response surface methodology. AEJ - Alexandria Engineering Journal. 2016;**55**:1201-1209. DOI: 10.1016/j.aej.2016.04.006

[84] Tomar M, Dewal H, Sonthalia A, Kumar N. Optimization of spark-ignition engine characteristics fuelled with oxygenated bio-additive (triacetin) using response surface methodology. Proceedings of the Institution of Mechanical Engineers, Part E: Journal of Process Mechanical Engineering. 2020;**235**:841-856. DOI: 10.1177/0954408920971110

[85] Najafi G, Ghobadian B, Yusaf TF, Safieddin Ardebili S, Mamat R. Optimization of performance and exhaust emission parameters of a SI (spark ignition) engine with gasoline–ethanol blended fuels using response surface methodology. Energy. 2015;**90**:1815-1829. DOI: 10.1016/j.energy.2015.07.004

[86] Vino V, Solomon R, Sreenath S. Performance analysis of spark ignition engine fueled with methanol/petrol fuel blends. Indian Journal of Science and Technology. 2013;**6**:4579-4582. DOI: 10.17485/ijst/2013/v6isp5.2

Chapter 3

Perspective Chapter: Methanol as a Fuel for Direct Methanol Fuel Cells (DMFCs) – Principles and Performance

Marcello Romagnoli and Veronica Testa

Abstract

Methanol, also known as methyl alcohol (CH_3OH), is a colorless, flammable, and volatile liquid produced commercially through the catalytic reaction of carbon monoxide and hydrogen or by gasification. Despite toxicity and serious health effects, methanol has recently gained attention as a feedstock for chemical synthesis, a solvent in industrial processes, an antifreeze agent, a potential solution for sustainable energy production, and as a potential alternative fuel for biofuel in automotive diesel engines in diesel vehicle applications. This is attributed to its notable energy density and convenient manageability when contrasted with hydrogen, a fuel more commonly employed in various other types of fuel cells. Proper handling and safety precautions are necessary when employing methanol as a fuel in direct methanol fuel cells (DMFCs) in portable electronic devices, backup power systems, and remote power generation applications. The performance of DMFCs is largely determined by the efficiency of the anode and cathode reactions, as well as the conductivity of the electrolyte. In the quest for more environmentally friendly and sustainable options, the uses of methanol are undergoing dynamic advancements, providing solutions that address both current energy demands and overarching environmental objectives.

Keywords: fuel cell, renewable energy, portable power, chemical conversion, methanol

1. Introduction

1.1 Fuel cell overview and environmental impact

Fuel cells are among the most promising systems for the production of electricity, with a view to replacing fossil fuels and reducing emissions. A fuel cell is an electrochemical device that converts the chemical energy of a fuel (generally hydrogen, methanol, or methane) and an oxidant (oxygen or air) into electrical energy. The main components of a fuel cell are two porous electrodes, anode (negative) and cathode (positive), and an electrolyte (**Figure 1**). Fuel cells operate based on the

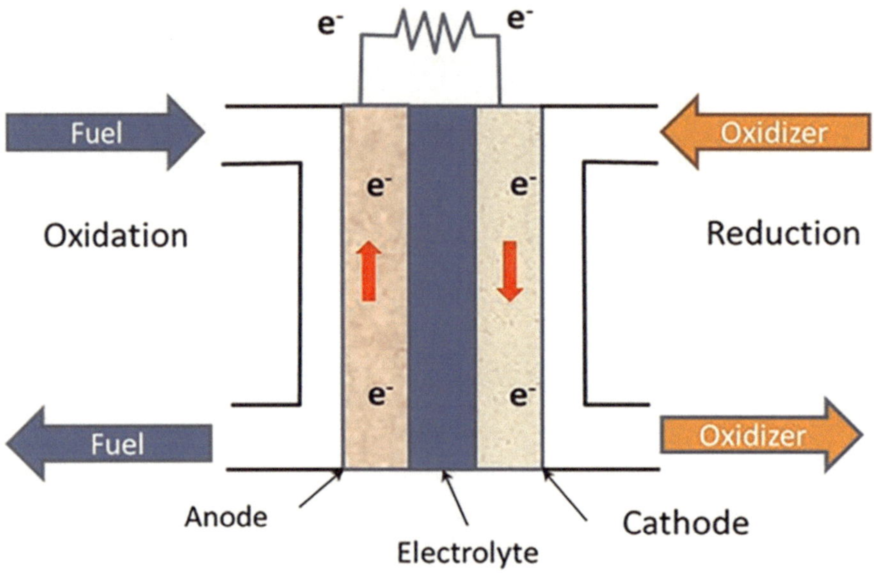

Figure 1.
Fuel cell schematic layout.

principle of reverse electrolysis: electrodes act as catalytic sites for cell reactions that basically consume hydrogen and oxygen, with water production and electric current passage in the external circuit. The electrolyte has the function of conducting the ions produced by one reaction and consumed by the other, closing the electrical circuit inside the cell [1]. At the anode, the fuel is oxidized, releasing electrons and producing protons as byproducts. The electrons flow through an external circuit, generating electrical power, and then return to the cathode where the oxidant is reduced.

A single cell normally produces a voltage of about 0.6, 7 V and currents between 300 and 800 mA/cm^2, so to obtain the desired power and voltage, multiple cells are arranged in series, by means of bipolar plates, to form the so-called "stack" [2]. The stacks are assembled in modules, to obtain generators of the required power in industrial applications. Potential applications include transportation (cars, busses, trucks, and trains), portable devices such as laptops or smartphones, stationary power generation, backup power for buildings, military and marine applications, and aerospace equipment. Fuel cells have several advantages over conventional power sources, such as internal combustion engines (**Figure 2**). They are highly efficient, with some types of fuel cells achieving efficiencies of up to 60% or more. They also emit no harmful pollutants, with water being the only byproduct of the reaction. Despite ongoing technical and economic challenges, fuel cell technology is expected to continue to advance and become more widely adopted in the coming years [3].

The widespread use of fuel cells has several environmental benefits that can contribute to the reduction of environmental risks compared to conventional gasoline systems and hybrids, through improved efficiency [4]. First, fuel cells produce electricity with high efficiency, resulting in less waste heat compared to combustion engines. This reduces the energy lost as heat, leading to lower greenhouse gas emissions and a reduced carbon footprint. Therefore, fuel cells can help mitigate the risks of climate change. Second, fuel cells produce minimal emissions of pollutants, such as carbon dioxide (CO_2), nitrogen oxides (NO_x), carbon monoxide (CO), sulfur oxides

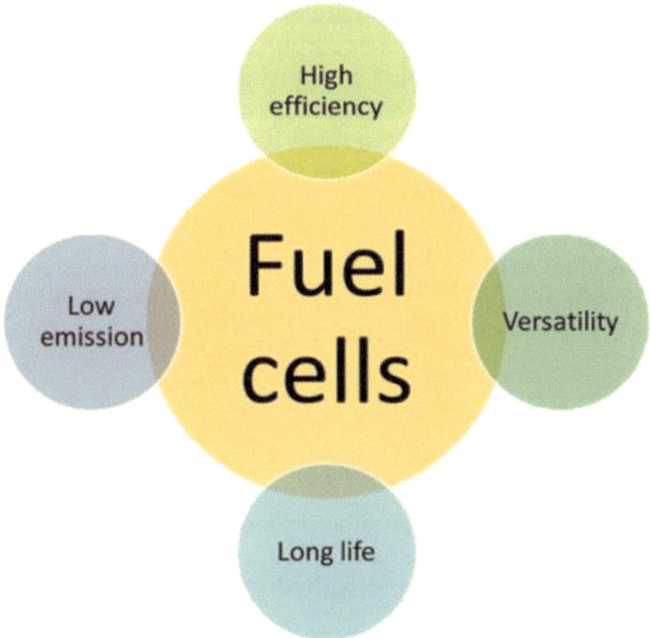

Figure 2.
Fuel cell advantages.

(SO_x), and particulate matter (PM), especially when using hydrogen as a fuel since the only byproduct is water [5]. Even when using other fuels such as methanol, natural gas, gasoline, or diesel, fuel cells produce lower emissions from vehicles compared to conventional combustion technologies (**Figure 3**): reduced carbon

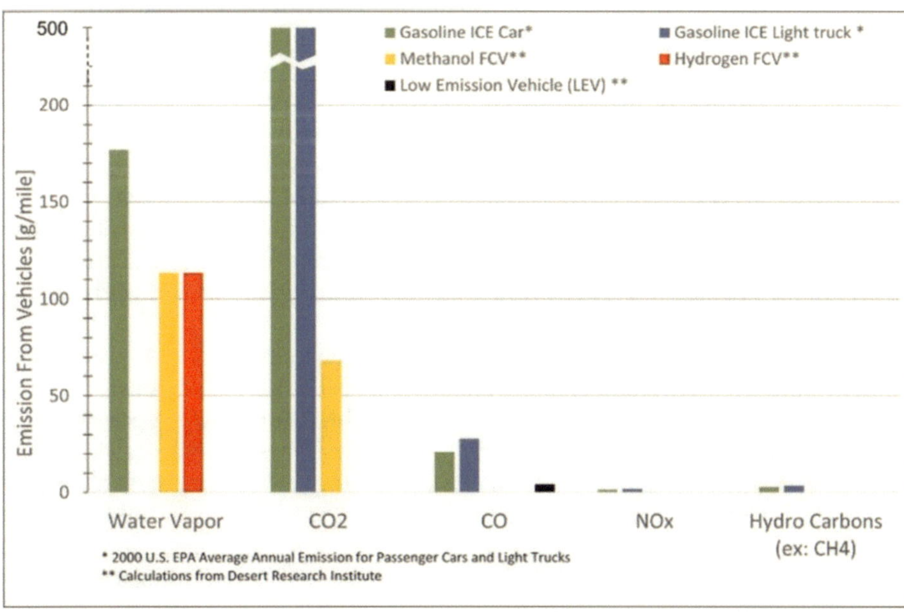

Figure 3.
Environmental impact of emissions from vehicles.

emissions are crucial to developed countries or regions in their efforts to meet target greenhouse gas emissions or reduce their carbon footprints [6].

As a result, fuel cells can contribute to reducing air pollution and associated environmental risks. Pollution is the most dangerous risk for the life on the earth. Fuel cells can help reduce the dependence on fossil fuels and associated environmental risks such as oil spills and air pollution from refineries. This is achieved by using renewable fuels such as hydrogen produced from renewable sources [7]. Fuel cells are cleaner, more efficient, and better suited for applications where clean, reliable power is required. Finally, fuel cells facilitate the development of a circular economy [8], by efficiently using waste streams. They can convert biogas from landfills or wastewater treatment plants into electricity, reducing waste disposal while generating useful energy [9].

2. Methanol fuel cell

Methanol fuel cells operate on the same principle as other types of fuel cells. However, they use methanol as a fuel instead of hydrogen and a polymer electrolyte membrane (PEM) to conduct the protons (hydrogen ions) that are generated during the electrochemical reaction.

2.1 Working principle

In a methanol fuel cell, the methanol is oxidized at the anode in the presence of a platinum-ruthenium (Pt-Ru) catalyst. This results in the production of hydrogen ions, electrons, carbon dioxide, and water. The hydrogen ions are then transported through sulfonated tetrafluoroethylene-based fluoropolymer-copolymer (which is most well known by the commercial name "Nafion®") to the cathode, where they combine with oxygen and electrons to produce water, releasing electrical energy that can be used to power devices [10]. The electrons are forced to flow through an external circuit, generating electric current, while the PEM prevents the direct mixing of the fuel and oxidant [1, 11]. The use of methanol as a fuel in fuel cells offers several advantages over other fuels such as hydrogen. Methanol is a liquid at room temperature, making it easier to transport, handling, and store. Methanol is also less expensive than hydrogen and can be produced from renewable energy sources or biomass. Besides, the high-energy density and the methanol's environmentally friendly property make methanol an attractive, safer, and less polluting alternative to other fuels [12]. Additionally, the use of hydrogen requires careful handling and storage, which can add to the overall system cost, and a longer start-up time.

Methanol fuel cells are still in the early stages of development, and researchers are working to improve the efficiency of methanol fuel cells and reduce their manufacturing cost [13]. One area of research is in developing new types of catalysts that are more efficient and less expensive than platinum-based catalysts. Another area of research is in developing new methods for producing and storing methanol fuel that are more sustainable and environmentally friendly. They are well suited for portable applications, such as powering electronic devices and small vehicles. They can also be used as backup power for critical infrastructure, such as telecommunications equipment and emergency response systems [14].

Methanol fuel cells can use methanol as fuel directly or indirectly. In an indirect system, methanol is first converted to hydrogen and carbon dioxide through a

	Electrical efficiency [%]	Operating temperature [°]
DMFC	20–30	70–90
IMFC	35–50	70–90 Low Temperature Polymer Electrolyte Membrane (LT-PEM) 160–200 High Temperature Polymer Electrolyte Membrane (HT-PEM)

Table 1.
Electrical efficiency and operating temperature of direct and indirect methanol fuel cells.

catalytic process called methanol reforming. Overall, the choice between a direct or indirect methanol fuel cell depends on the specific application and the trade-offs among simplicity of system design, electrical efficiency, cost, and operating temperature (**Table 1**).

2.2 Direct methanol fuel cells (DMFCs)

Direct methanol fuel cells (DMFCs) are attractive for portable applications due to their high density but most research works have focused on fundamental issues of internal components, such as membranes and catalysts. The direct generation of electricity from methanol and oxygen supplied from external tanks simplifies the system design without the need for an external reformer to convert the fuel into hydrogen.

Figure 4 shows a schematic representation of DMFC, in which the overall chemical reaction involved is given in Eq. (1).

$$2\,CH_3OH + 3O_2 \rightarrow 4\,H_2O + 2\,CO_2 \qquad E^0 = 1.21\,V \tag{1}$$

The overall chemical reaction can be divided into two half-reactions. At the anode, oxidation occurs, as given in Eq. (2).

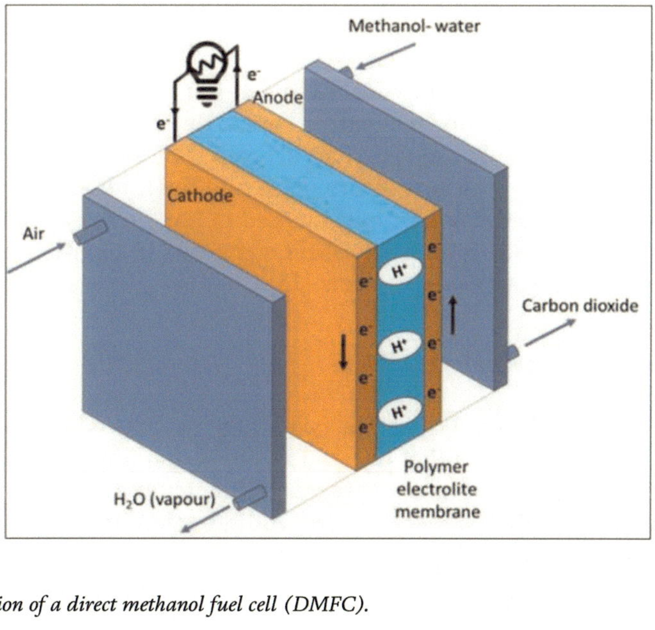

Figure 4.
Schematic illustration of a direct methanol fuel cell (DMFC).

$$2CH_3OH + 2H_2O \rightarrow 12H^+ + 12e^- + 2CO_2 \tag{2}$$

While at the cathode, reduction takes place, as given in Eq. (3).

$$O_2 + 12H^+ + 12e^- \rightarrow 6H_2O \tag{3}$$

Some of the advantages of a DMFC include:

- Simple and compact system design.
- Few minutes for start-up time.
- Widely used and well established for low-power applications.
- High-energy density.
- Low emission of pollutants.

The methanol must first diffuse through the anode to reach the catalyst layer, where it is oxidized to produce protons and electrons. The protons then pass through the membrane to the cathode, where they react with oxygen to produce water. The time required for the methanol concentration at the anode to reach the optimal level can affect the start-up time of the DMFC. It means that the start-up time can vary depending on several factors, such as the fuel cell design, the methanol concentration, and the operating temperature. However, a DMFC requires high purity methanol, and the membrane can become susceptible to damage when exposed to subzero temperatures due to the formation of ice crystals, which presents challenges for methanol storage. Finally, direct methanol fuel cells can be classified into two primary types: passive and active, which are different in terms of the power output and components. Passive fuel cells generate power through the natural flow of liquid methanol through the cell, while active fuel cells use a pump to circulate the methanol to the anode. The pump is used to regulate the methanol flow rate, ensuring that the fuel is delivered to the anode at a consistent rate.

The control system monitors the fuel cell and adjusts the pump speed and methanol concentration to maintain optimal performance. In this way, active fuel cells can achieve higher power densities than passive fuel cells. The choice of which type to use depends on the specific application requirements and the trade-off between performance and cost. Passive DMFCs are typically used in low-power applications such as portable electronic devices, while active DMFCs are used in high-power applications such as vehicles and backup power systems [15].

2.3 Indirect methanol fuel cells (IMFCs)

As described above, the employed fuel is hydrogen produced by methanol process in a reformer system (**Figure 5**) that decomposes methanol into hydrogen and carbon dioxide. Afterward, the hydrogen is purified and directed to the anode of the fuel cell, whereas the carbon dioxide is either recycled or discharged. In Eq. (4), the overall electrochemical reaction is explained.

$$CH_3OH + H_2O \rightarrow 3\,H_2 + CO_2\ (Reformer)\ 2\,H_2 + O_2 \rightarrow 2\,H_2O \tag{4}$$

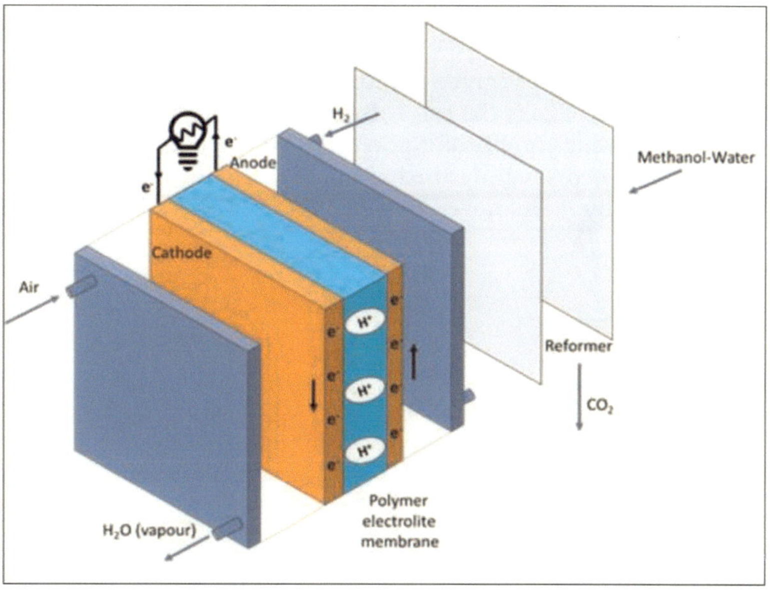

Figure 5.
Schematic illustration of indirect methanol fuel cell (IMFC).

The main advantage of the IMFC is that the use of hydrogen can increase the efficiency (see **Table 1**) of the fuel cell, compared to a direct system.

The high efficiency and the possibility of cold storage, as there is no contact with the membrane, are the reasons why IMFCs are widely used, despite more complex system design and higher costs. The start-up time of an IMFC can be affected by various factors, such as the reformer design, the hydrogen concentration, and the operating temperature. Since hydrogen diffuses faster than methanol, it reaches the catalyst layer more quickly, leading to a faster start-up time. During start-up, the hydrogen concentration at the anode is relatively high, which allows for more rapid oxidation and proton production [15].

3. Performance of direct methanol fuel cells

Since direct methanol fuel cells are not the only type, or even the most well known, it is common for their performance to be compared to that of other fuel cells, especially those that use hydrogen (H_2). Some of the parameters considered include energy density, lifetime, environmental impact, capital expenditure (CapEx) and operational expenditure (OpEx) cost, and flexibility of use, among others.

3.1 Efficiency of methanol fuel cells (MFCs)

The thermodynamic theoretical energy conversion efficiency (η_{th}) is defined as the ratio between the free energy (ΔG) and the chemical energy (enthalpy ΔH) of the fuel cell (Eq. (5)).

$$\eta_{th} = \frac{\Delta G}{\Delta H} \tag{5}$$

The η_{th} value of a direct methanol fuel cell (DMFC) at 25°C reaches 97% [16], when the η_{th} value for the H_2 stops at 83% [17]. The practical energy efficiency, however, is much lower. For the DMFC, it is given by Eq. (6). The voltaic efficiency $\eta_{voltaic}$ is defined by Eq. (7), where V_{cell} is the cell voltage at an operating current density of I and ΔE the thermodynamic equilibrium potential. The η_{fuel}, in a DMFC, is fuel efficiency due to methanol crossover defined as Eq. (8), where $I_{x,\ over}$ is an equivalent current density caused by methanol crossover under the operating current density of I [18].

$$\eta = \eta_{th}\eta_{voltaic}\eta_{fuel} \tag{6}$$

$$\eta_{voltaic} = \frac{V_{cell}}{\Delta E} \tag{7}$$

$$\eta_{fuel} = \frac{I}{I + I_{x,over}} \tag{8}$$

The phenomenon will be described in more detail later in Chapter 4. Based on these reasons, the energy conversion efficiency achievable in current operational cells ranges from 20 to 30% [19].

3.2 Power output of methanol fuel cells

The thermodynamic equilibrium cell potential for a DMFC is 1.21 V, which is higher than the actual open circuit voltage, typically lower than 0.9 V, mainly due to the factors such as the irreversible adsorption of methanol-derived intermediate species at the anode and fuel crossover [18, 20]. This last issue arises due to the permeability of current perfluorosulfonic acid membranes, which function as the electrolyte in DMFCs, to methanol. As a consequence, fuel crossover occurs, resulting in reduced performance and limiting the utilization of aqueous mixtures with high methanol concentrations. The primary reason behind this limitation is the easy miscibility of methanol with water, causing it to diffuse into the water that forms an integral part of the electrolyte structure. The methanol that reaches the air cathode, which has a platinum catalyst, is oxidated. The reaction is a waste of fuel and the loss of methanol is often expressed as a "crossover current": the current equivalent to that which would be produced by the methanol if it had reacted properly on the anode [1]. As current flows out of cells, the closed-circuit voltage is further reduced and changes with variations in current density. This behavior is graphically depicted in the polarization curve for DMFC, illustrated in **Figure 6**—DMFC polarization curve. It serves to demonstrate the relationship between cell voltage and current density, facilitating the analysis of DMFC's performance, including its inherent limitations. Notably, the polarization curve exhibits higher voltage values at lower current densities, gradually diminishing as the current increases. Typically, it can be classified into three distinct regions: activation, ohmic, and mass transport. In the activation region, characterized by low current density, the voltage decreases as the current density increases. This phenomenon can be attributed to the sluggish kinetics of methanol oxidation at the anode and oxygen reduction at the cathode. The anode reaction exhibits poor electrode kinetics, especially at lower temperatures. Consequently, the utilization of enhanced catalysts and higher operating temperatures becomes necessary to improve the reaction kinetics. The cathode reaction, involving oxygen reduction, also suffers from sluggish kinetics, although it is comparatively less problematic than with

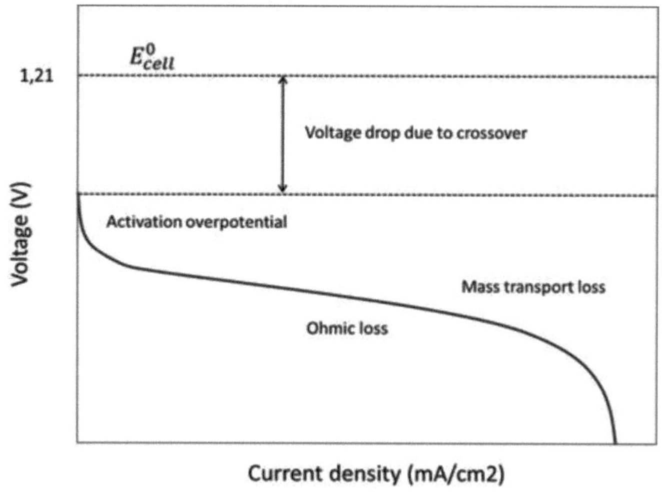

Figure 6.
Direct methanol fuel cell (DMFC) polarization curve.

aqueous mineral acid electrolytes. This voltage loss due to sluggish kinetics is second only to that caused by crossover [21].

The intermediate region, characterized by moderate current densities, is known as the ohmic control region. In this interval, the voltage exhibits an almost linear decrease. This behavior is primarily attributed to the electrical resistances encountered within the fuel cell, including the electrodes, electrolyte, and cell connections. The voltage drop in this region follows Ohm's law, which relates the voltage to the current and resistance within the system. In the last region, where current densities are high, mass transport resistance is the dominant factor. As a result, the cell's potential decreases rapidly when the concentration of one of the reactants approaches zero at the corresponding catalyst layer. This behavior is not unique to DMFCs but is inherent to all fuel cells, albeit with varying values, as the underlying phenomena influencing the potential curve are universally present. Nevertheless, modern DMFCs have demonstrated notable performance with power densities surpassing 300 $mWcm^{-2}$ when utilizing oxygen and 200 $mWcm^{-2}$ in air under pressure. It is important to note that these power densities are considerably lower than those achieved by hydrogen-fueled fuel cells. Despite these challenges, DMFCs hold the potential to be highly cost-effective and competitive with internal combustion engines in various applications [21].

3.3 Comparison with other fuel cell types

Direct methanol fuel cells offer several advantages compared to other types of fuel cells, including ease of fuel delivery and storage, low cost of methanol, operation at low temperature and pressure, absence of humidification requirements, and reduced design complexity (**Table 2**).

Methanol, being a liquid with properties like conventional fuels, can be stored in tanks at ambient pressure, resembling the setup of a gasoline tank. Refueling a DMFC vehicle is comparable to traditional methods in terms of time and procedure. Unlike hydrogen, methanol does not require high-pressure storage tanks, reducing the need for energy-intensive compression. Methanol can be produced through synthesis,

Fuel	Fuel cell type	Thermodynamic theoretical energy conversion efficiency (H_{TH})	Real energy conversion efficiency (H_{TH})	Operating temperature (°K)	Applications
Hydrogen	PEMFC	83%	40–60%	325–355 [22]	Backup power Portable power Distributed generation Transportation Specialty vehicles
	Alkaline Fuel Cell (AFC)		60%	325–475 [22]	Transportation, space travel
	Phosphate Acid Fuel Cell (PAFC)		40%	465–485 [22]	Combined Heat and Power (CHP), power plants
	Molten Carbonate Fuel Cell (MCFC)		50%	875–925 [22]	CHP, power plants
	SOFC		50%	875–1275 [22]	CHP, power plants
Methanol	DMFC	97%	20–30%	355–475 [22]	Transport, mobile equipments

Table 2.
Fuel cells comparison chart.

involving the reaction between CO_2 and H_2, or through biosynthesis, involving the reaction between CH_4 and O_2. The reaction taking place within the DMFC generates CO_2 and H_2O as byproducts. When produced via synthesis, DMFCs do not contribute to the increase of carbon dioxide in the atmosphere. Unlike polymer electrolyte membrane fuel cells (PEMFCs) or solid oxide fuel cells (SOFCs), DMFCs do not require external or internal reforming processes, as they can utilize the organic compound directly. The challenges include the lower efficiency and power density, as well as the higher cost of DMFCs compared to H_2-based fuel cells.

4. Challenges and solutions

4.1 Methanol crossover

As mentioned earlier, the phenomenon of methanol crossover refers to its transfer from the anode to the cathode, through the electrolyte membrane. At the cathode, it comes into direct contact with atmospheric oxygen and undergoes oxidation. This process leads to several disadvantages, including a decrease in cell voltage, lower current density compared to the maximum theoretical value, and fuel consumption without corresponding electricity production. It contributes to the fact that only approximately 30% of the energy that can be released from methanol, based on the aforementioned reactions, can be converted into electrical energy [23, 24]. Multiple physical and chemical factors exert influence over methanol crossover [25]. Among the former, noteworthy physical parameters include temperature, pressure, membrane thickness, and current density. As for the chemical parameters, the concentration of methanol in the anode plays a significant role: it increases with increase in the amount of methanol fraction at the anode and the current generated by the cell [26]. Numerous studies have demonstrated that an elevated temperature leads to an increased crossover phenomenon. However, the overall impact on the cell's total efficiency due to temperature variation is intricate. In reality, crossover is a composite result of diffusion and proton drag. Diffusion tends to diminish as current density rises, whereas an escalation in proton drags results in a decline in the methanol concentration at the anode. Conversely, if the methanol concentration is maintained constant at the anode/membrane interface, proton drag intensifies as a greater amount of protons is generated through oxidation reactions [25].

4.2 Catalyst poisoning

Catalyst poisoning poses a significant challenge for platinum group metal catalysts utilized in DMFCs. During the alcohol oxidation process, carbon monoxide (CO) molecules, generated as reaction intermediates, can hinder the reaction by obstructing the active sites. This problem can be mitigated by using binary catalysts, i.e., Pt-Ru.

Additionally, the accumulation of carbon dioxide on the cathodic surface can lead to the formation of bubbles, blocking the flow channels and reducing the effective surface area of the electrode, thereby limiting the performance of the fuel cell [27].

Although platinum (Pt) is commonly employed as a catalyst in DMFCs, it is prone to catalyst poisoning caused by carbonaceous species, resulting in reduced DMFC performance.

The detrimental effects of methanol poisoning at the cathode can be mitigated by increasing the oxygen partial pressure, which promotes the oxidation of carbon monoxide to carbon dioxide. Additionally, this increased oxygen stoichiometry helps to counteract the adsorption of methanol and facilitates the physical removal of the aqueous solution of alcohol that permeates through the membrane, along with the water formed by the reaction. This preventive measure helps prevent electrode flooding. Compared to other fuel cells, DMFCs are more susceptible to cathode flooding due to the significant contribution of liquid water and methanol to the anode [28]. The use of binary or ternary catalysts in DMFC has also been studied in several research articles. These catalysts are commonly used due to their high catalytic activity for methanol oxidation reaction (MOR) [29, 30].

Pt/Ru is a binary catalyst that has been shown to have high activity and stability for MOR in DMFCs. Moreover, the addition of Ru can improve the catalyst performance and durability by reducing the carbon monoxide poisoning effect and increasing the tolerance to methanol crossover [31]. Also, the cost of the fuel cell can be reduced. Optimizing the synthesis method, particle size, and support material can improve the performances of Pt/Ru catalysts [32]. Other binary and ternary catalysts, such as platinum/palladium (Pt/Pd) and Pt/Pd/Ru, have also been studied for their potential use in DMFCs [33, 34].

4.3 Electrode degradation

In addition to catalyst poisoning, other forms of catalyst deterioration can occur, causing more general electrode degradation. These include: dissolution and agglomeration [35].

In the specific working conditions experienced by DMFCs, conventional Pt-Ru catalysts can undergo dissolution on the anode side. The migration and accumulation of these elements at the cathode, as they pass through the electrolyte, have been observed. This phenomenon leads to a decrease in catalyst activity in both electrodes, resulting in reduced overall cell performance [36].

The catalyst particles used in DMFCs are of nanoscale dimensions to maximize the surface area per unit mass. This is crucial because catalytic reactions occur on the surface of these particles. However, due to various factors, these particles can agglomerate, increasing their size but reducing the total available surface area, thereby diminishing the cell's efficiency [37].

As seen in the previous paragraph, methanol crossover can cause catalyst poisoning and cathode degradation, resulting in reduced performance. Finally, during the electrochemical reactions in the DMFC, carbon dioxide can be produced at the anode. The accumulation of carbon dioxide can lead to the formation of bubbles, blocking the flow channels and reducing the active surface area of the electrode, thus limiting the fuel cell's performance [27, 38].

4.4 Membrane degradation

The membrane, which separates the anode and cathode compartments in a DMFC, can also undergo degradation. Exposure to methanol, high temperatures, and other operating conditions can cause membrane swelling, loss of mechanical strength, or chemical degradation. Membrane degradation can lead to increased methanol crossover, decreased proton conductivity, and reduced overall performance of the fuel cell. Several authors have observed that high methanol concentration leads to the thinning of the polymer membrane, which serves as the electrolyte, due to its solubility in alcohol. As the thickness of the electrolyte decreases, the crossover of methanol increases. Therefore, a high percentage of methanol introduced at the anode ultimately contributes to the decrease in cell performance. At elevated operating temperatures, the degradation rate of the membrane accelerates due to the development of pinholes in the membrane, combined with cathode degradation and delamination. Another cause of degradation is attributed to the stack assembly process. Specifically, the tightening process can cause damage due to the mechanical stresses exerted on the membrane. Among the strategies to mitigate this issue, various approaches can be considered: investigating different cross-linking procedures, incorporating reinforcements or nanoscale inorganic fillers into the membrane, and exploring alternative materials for constructing the electrolyte [35].

4.5 Strategies for improving performance

The main strategies to improve the performance of DMFCs involve addressing the issues mentioned earlier. By mitigating these challenges, significant enhancements can be achieved. Reducing the crossover phenomenon is one of the primary research directions pursued by various scientists.

One strategy is to use membranes that are less permeable to methanol, such as those using SPEEK (sulfonated polyether ether ketone), instead of the well-established sulfonated tetrafluoroethylene polymers [39, 40]. Enhancing the efficiency of the catalyst at the anode of the cell enables the improvement of methanol oxidation, thereby reducing the fuel permeation across the electrolyte membrane. Formulations incorporating Pt-Ru have exhibited promising efficiency. Other ongoing investigations involve: catalyst particles with nanometric or even atomic-scale dimensions, as well as novel element alloys and supported materials [41, 42]; and adding ZrO_2 or a thin layer of palladium to the membrane. These methods are capable of reducing methanol crossover to a certain extent, but none can prevent methanol crossover completely [25].

Another approach is to operate at low fuel concentrations, thereby decreasing the driving force for methanol crossover [43]. It is crucial to acknowledge that a comprehensive reduction of crossover in DMFCs often requires a combination of these methods. This is because each approach focuses on distinct aspects of the crossover phenomenon. To mitigate electrode degradation in DMFCs, researchers are exploring various strategies, such as developing more efficient catalyst materials, improving membrane durability, and optimizing system designs to minimize methanol crossover and carbon dioxide accumulation. Ongoing research aims to enhance the long-term stability and performance of DMFC electrodes for their successful commercialization.

4.6 DMFC massive production aspects

A massive adoption of DMFC, in addition to addressing previously observed issues, necessitates a reduction in cell costs. The combined expense of the membrane electrode assembly (MEA) and bipolar plates constitutes just over 50% of the overall cost, whereas the cost associated with the balance of plant (BoP) is comparatively lower. Attention should be focused on these two former components to explore possibilities for reducing production costs [44]. Maximizing automation in the production process can facilitate cost reduction without compromising on product quality. A feasible manufacturing facility should apply catalytic inks on both the anodic and cathodic sides. Discontinuous methods such as slot-die coating have been associated with issues such as layer inhomogeneity, local delamination, recurring defects, and raw material waste. The slot-die coating technique is seen as having the potential to partially mitigate these problems [45–47]. It is utilized to apply the catalytic ink on the membrane, followed by the application and pressing of the gas diffusion layer (GDL), either through hot or through cold methods, to obtain the MEA. However, it is not the only technique under consideration. Other technologies, such as: tape casting; gravure printing [46]; bar coating; screen printing [48], air spraying, ultrasonic spray deposition, and inkjet printing, have been explored as promising methods for achieving continuous production [49]. In terms of bipolar plates, improvements can be achieved by pressing graphite powders with binders or by pressing sheets of steel or aluminum alloys [50]. Finally, the introduction of a high degree of automation on production lines is the third essential element. This would enable increased production quantities without compromising on product quality's variability. The automation lines should encompass MEA and bipolar plate production technologies, while also being capable of assembling stacks.

5. Applications of methanol fuel cells

5.1 Portable electronic devices

One prospective domain where direct methanol fuel cells (DMFCs) can be utilized is the provision of power to electrical and electronic devices. Several examples include charging mobile phones and laptops, supplying energy to portable medical devices employed in field hospitals, and powering motor homes. Additionally, DMFCs can serve as auxiliary power units (APUs) for weather stations and radio repeaters located in areas lacking power infrastructure. DMFCs exhibit great potential for these applications due to their advantageous attributes, including high-energy density, lightweight design, compactness, simplicity, and convenient and rapid recharging capabilities [51].

In comparison to conventional lithium-ion rechargeable batteries, DMFCs offer distinct advantages. First, they possess a theoretically higher specific energy density of 6000 Wh/kg, surpassing the best rechargeable battery systems with a theoretical energy density of 600 Wh/kg [28]. This implies that DMFCs can store more energy per unit of weight. Additionally, they exhibit a significantly shorter charging time. While it takes hours to recharge batteries, the tank connected to the methanol cell's anode can be refilled in just a few minutes without interrupting the cell's electricity output. This rapid refueling capability sets DMFCs apart from batteries. Compared to PEMFCs, direct methanol fuel cells (DMFCs) have the advantage of not requiring the storage of hydrogen under pressure or in the form of hydrides. The latter can be replenished with H_2 but necessitate not only the availability of hydrogen gas but also a specific procedure. Ultimately, the average user may perceive hydrogen-based solutions as more hazardous and complex compared to those utilizing methanol. Regarding the hazardous nature of methanol, it is important to note that its level of danger is not significantly different from those of other conventional fuels such as gasoline, diesel, natural gas, or liquefied petroleum gas (LPG). There are already DMFC-based products available on the market, and there is a strong interest in developing new, more efficient, cost-effective, and long-lasting solutions [52].

5.2 Electric vehicles

An attempted indirect use of methanol involved combining it with a PEMFC through a reformer capable of converting alcohol into H_2 and CO_2. This approach appeared to be highly promising in the 1990s and early 2000s. Several major global vehicle manufacturers pursued this path but had to abandon it due to significant technical and economic challenges [51].

Material handling is another area of use for DMFCs. Forklifts find extensive application in logistical, industrial, and construction sectors. They are propelled by either electric motors or internal combustion engines, constituting the two primary propulsion methods. In the case of electric forklifts, power can be sourced from batteries or generated through fuel cells. Notably, in 2020, the number of electric forklift shipments exceeded twice the number of shipments for internal combustion engine (ICE) units. There are several advantages offered by a fuel cell electric system. The main advantage lies in the comparable charging times with traditional internal combustion engine (ICE) forklifts. Furthermore, in contrast to battery-powered electric forklifts, fuel cell systems maintain maximum operational efficiency as long as there is fuel in the tanks, without experiencing performance degradation due to battery charge depletion. Forklifts utilizing proton exchange membrane (PEM) fuel cells are the most prevalent in the market. However, they require infrastructure capable of supplying pressurized H_2, which needs to be stored in specifically designed areas with significant safety systems. Hydrogen can be purchased or internally generated through electrolyzers. In contrast, when using DMFCs, the supply of methanol poses fewer issues as it is a liquid similar to traditional fuel. It does not need to be stored under high pressure in dedicated tanks. Of course, all necessary precautions must be taken for handling a flammable and toxic fuel. Additionally, methanol cannot be produced

Characteristic	Efoy pro 12,000 duo [57]	Blue world [58]
Nominal voltage	24 VDC/48 VDC	38 to 60 VDC
Power output	500 W	5 kW
Charging current	20.83 A/10.42 A	
Weight	32.0 kg	530 kg
Warranty (operating hour)	3000	10,000
Nominal consumption	0.9 l/kWh	0.55 l/kWh
Dimensions D × W × H	640 × 441 × 310 mm	845 × 1433 × 1354 mm
Operating temperature	−20 to +50°C	−10 to +50°C

Table 3.
Technical specifications of some commercial DMFC power generators.

on-site, as is the case with H_2 when, for example, an electrolyzer is connected to a renewable energy source such as photovoltaic or wind power [53–55].

5.3 Power generation in remote areas

In remote areas, even where electricity is not available, it may be necessary to have DMFC to power systems such as radio repeaters and the Internet network; remote monitoring and land survey systems; navigation systems, etc.

In these applications, it is much more convenient to provide reliable off-grid energy via fuel cells than to establish an expensive and difficult grid connection or use photovoltaic panels that have insurmountable limitations when placed in areas with unfavorable weather conditions.

The use of classical portable batteries that provide the full amount of energy is problematic due to their need to be replaced or recharged. DMFC systems can provide the full amount of energy by combining them with a smaller amount of batteries capable of providing the energy to overcome peak demands [56].

There are DMFC-based systems on the market which are capable of delivering power that is not very high, but sufficient for various applications used in remote areas. Some characteristics are listed in **Table 3**.

Even in cases where the system is installed in remote areas, the operator has the ability to confirm the operational condition of the system by establishing a connection to the cloud server via a mobile phone network. Furthermore, it is worth noting that methanol aqueous solution is a safe substance that can be stored for extended periods without undergoing any degradation.

5.4 Potential applications

Direct methanol fuel cells can have interesting developments, especially if they will be able to reduce the disadvantages that have been described above. This would allow a higher power density and open up new applications. For example, it would allow them to be used for small electric cars, even as range extenders for battery electric vehicles (BEVs). Solutions have been proposed for small (Class III) forklifts for DMFC-powered mobile pick-ups. From a search on the Internet, however, the company proposing the methanol system does not appear to have a website, so it is not possible to say whether they are still operational on the market [59–61].

Nevertheless, there are other companies offering DMFCs of comparable or even higher power. In particular, one of them also offers its systems for transport applications such as cars, trucks, and maritime transport. Methanol fuel cells are combined with batteries in hybrid configurations. The battery is very good for short high-power needs and the fuel cell for an extended need of energy [62].

While not directly related to DMFCs, it is worth mentioning alternative solutions that utilize methanol as a convenient hydrogen storage and transport system. In these systems, methanol is supplied to extract hydrogen for utilization in cells such as PEMFCs.

6. Methanol fuel cells for future

6.1 Advances in methanol fuel cell technology

As described above, DMFCs suffer from the slow kinetics of methanol oxidation reactions. To reduce this problem, expensive Pt-based catalysts are introduced at both the anode and cathode, making them unsuitable for large-scale commercialization. The noble metal loading of up to 4 mg cm^{-2} on the anode is too much expensive for a lot of applications. The anode catalyst activity has to be increased by a factor of at least 10 to be able to reduce the noble metal loading to a more reasonable 0.5 mg cm^{-2} [63].

Another line of solution is to develop lower Pt and non-Pt catalysts. Low-Pt alloys and core-shell-like catalysts have yielded important results. The latter are nanomaterials that have a structure consisting of a central part, the inner core material, surrounded by a shell material. They can provide performance superior to that of the generally applied conventional catalysts [64, 65]. Heat-treated Pd-Me, Ru-Se, and MeNxCy-based Pt-free catalysts are also being investigated. Methanol crossover limitation is a significant challenge in DMFCs. Improving DMFC performance would be greatly beneficial, if either a methanol-impermeable electrolyte or a methanol-tolerant cathode is available. Extensive research has focused on exploring alternative membrane materials to minimize methanol crossover effects. However, current electrolyte materials face water management issues, which restrict their operation to temperatures below 100°C at ambient pressures. Increasing the operating temperature to 150° C would significantly enhance the kinetics of the anode reaction. This necessitates the development of new materials that exhibit high conductivity without the need for humidification [63].

6.2 Market prospects for methanol fuel cells

The DMFC market is moderately consolidated and is expected to show substantial growth during the period 2022–2027, owing to a strong demand for electronics coupled with the need for clean energy. In particular, the compounded average growth rate expected is more than 13% during this period of time. China, Japan, India, and other countries in Asia-Pacific are expected to dominate the direct methanol fuel cell market. North America's and European DMFC markets are poised for growth until 2030. The region benefits from capital-intensive companies and substantial investments in product launches. A positive economic outlook, along with focused marketing efforts, will bolster the market. The market for direct methanol fuel cells in Latin America is expected to exceed earlier growth projections, driven by strong domestic demand and a rise in export activities. However, inflation and high-interest rates may impact negatively. Middle East countries' focus on economic diversification drives strong demand for DMFCs. United Arab Emirates (UAE), Saudi Arabia, and other nations offer promising prospects, along with selected African countries, for companies in this sector [66]. The main industries that employ DMFCs include the electronics, automotive, and industrial sectors. DMFCs have diverse applications due to their higher energy density compared to traditional lithium-ion batteries and the great ease of storage of methanol compared to H_2. This particular feature is anticipated to make a significant contribution to the growth and advancement of the global market [67]. In order to overcome technical and economic limitations, there is an increase in Research & Development (R&D) investments that should lead to significant improvements. It's important to underline that while methanol fuel cells produce fewer emissions than traditional power sources, such as gasoline-powered generators, the production and use of methanol fuel can still have an environmental impact [68].

7. Conclusion

Methanol (CH_3OH) is the simplest organic alcohol, consisting of one carbon, four hydrogen, and one oxygen atom. It is a promising fuel for future energy systems due to its high-energy density, ease of production from renewable energy sources or biomass, and easy storage and transportability. Methanol is typically not only produced from fossil fuels, such as natural gas, which can contribute to greenhouse gas emissions, but it can also be produced from sustainable biomass, often called biomethanol, or from carbon dioxide and hydrogen produced from renewable electricity. In these last cases, methanol can be considered as environmentally friendly for the planet.

In comparison to other types of fuel cells, DMFCs offer various benefits, including their ability to function at low temperatures and their simplicity and compactness.

The fundamental working principle of a DMFC involves the electrochemical oxidation of methanol at the anode, resulting in the production of protons, electrons, and carbon dioxide. The protons flow through a proton exchange membrane to the cathode, while the electrons generate an electric current through an external circuit. At the cathode, airborne oxygen combines with the protons and electrons to produce water. One of the major advantages of DMFCs is their ability to utilize liquid fuel, making storage and transport simpler than other fuel cell types. Additionally, their simple design results in a more compact and lightweight system. However, DMFCs are limited by the lower energy density of methanol in comparison to hydrogen, which can impact their performance. Additionally, the overall energy efficiency of DMFCs is lower than that of other fuel cell types, such as polymer electrolyte membrane fuel cells (PEMFCs). Research efforts undertaken on DMFCs are focused on enhancing their efficiency, durability, and cost-effectiveness to improve their competitiveness as a power source. This includes exploring new catalyst materials, improving the design of the fuel cell components, and developing new techniques for producing and storing methanol fuel. For example, researchers are investigating the use of higher methanol concentrations and alternative materials, such as nanomaterials, as catalysts to improve DMFC performance. Additionally, efforts are underway to develop more effective and efficient methods for producing and storing methanol fuel. Despite the limitations of DMFCs, they offer several benefits over other fuel cell types and have the potential to be a promising technology for both portable and stationary power applications. Further research and development are necessary to fine-tune their performance and make them more competitive with other power sources.

Author details

Marcello Romagnoli* and Veronica Testa
Department of Engineering "Enzo Ferrari", Interdepartmental Center for Research and Services in the field of Production, Storage and Utilization of Hydrogen H2-MO.RE, University of Modena and Reggio Emilia, Modena, Italy

*Address all correspondence to: marcello.romagnoli@unimore.it

References

[1] Dicks AL, Rand DAJ. Fuel Cell Systems Explained. Hoboken, New Jersey, U.S: John Wiley & Sons; 2018. pp. 69-133. DOI: 10.1002/9781118706992

[2] Perry ML, Fuller TF. A historical perspective of fuel cell technology in the 20th century. Journal of the Electrochemical Society. 2002;**149**:S59. DOI: 10.1149/1.1488651

[3] Ulf B, Schönbein CF, GWR. The birth of the fuel cell 1835-1845. Lucerne, Switzerland: European Fuel Cell Forum; 2000

[4] McCarthy R, Yang C. Determining marginal electricity for near-term plug-in and fuel cell vehicle demands in California: Impacts on vehicle greenhouse gas emissions. Journal of Power Sources. 2010;**195**:2099-2109. DOI: 10.1016/j.jpowsour.2009.10.024

[5] Upadhyaya J, Peters RW, Fouad FH, Ahluwalia RK, Doss ED, Das T. Environmental impact of fuel cell technology for electric power generation: An overview and case studies. In: Proceeding in AIChE Annual Meeting, Austin, Texas; 2004

[6] Damm C, Strobach E, Robbins C, Broch A, Turner R, Hoekman SK. Development of the renewable energy deployment and display (REDD) facility at the Desert Research Institute-volume 2 of energy sustainability. In: ASME 2014 12th International Conference on Fuel Cell Science, Engineering and Technology, Jun 30-Jul 2, 2014. Boston, Massachusetts, USA: American Society of Mechanical Engineers; 2014. DOI: 10.1115/ES2014-6626

[7] Megía PJ, Vizcaíno AJ, Calles JA, Carrero A. Hydrogen production technologies: From fossil fuels toward renewable sources. A mini review. Energy & Fuels. 2021;**35**:16403-16415. DOI: 10.1021/acs.energyfuels.1c02501

[8] Ahmed AA, Nazzal MA, Darras BM, Deiab IM. A comprehensive sustainability assessment of battery electric vehicles, fuel cell electric vehicles, and internal combustion engine vehicles through a comparative circular economy assessment approach. Sustainability. 2022;**15**:171. DOI: 10.3390/su15010171

[9] Osman AI, Mehta N, Elgarahy AM, Hefny M, Al-Hinai A, Al-Muhtaseb AH, et al. Hydrogen production, storage, utilisation and environmental impacts: A review. Environmental Chemistry Letters. 2022;**20**:153-188. DOI: 10.1007/s10311-021-01322-8

[10] Heinzel A, Barragán VM. A review of the state-of-the-art of the methanol crossover in direct methanol fuel cells. Journal of Power Sources. 1999;**84**:70-74. DOI: 10.1016/S0378-7753(99)00302-X

[11] Shripad T, Revankar PM. Fuel Cells: Principles, Design, and Analysis. Boca Raton, Florida: CRC Press; 2014

[12] Schulze Lohoff A, Günther D, Hehemann M, Müller M, Stolten D. Extending the lifetime of direct methanol fuel cell systems to more than 20,000 h by applying ion exchange resins. International Journal of Hydrogen Energy. 2016;**41**:15325-15334. DOI: 10.1016/j.ijhydene.2016.06.207

[13] Barbera O, Stassi A, Sebastián D, Baglio V, Aricò AS. Direct Methanol Fuel Cell Stack Design and Test in the Framework of DURAMET Project. Vol. 93. Switzerland: Advances in Science and

Technology; 2014. pp. 65-69. DOI: 10.4028/www.scientific.net/AST.93.65

[14] Yuan Z, Zhang Y, Fu W, Li Z, Liu X. Investigation of a small-volume direct methanol fuel cell stack for portable applications. Energy. 2013;**51**: 462-467. DOI: 10.1016/j.energy.2012.12.033

[15] Kamarudin SK, Achmad F, Daud WRW. Overview on the application of direct methanol fuel cell (DMFC) for portable electronic devices. International Journal of Hydrogen Energy. 2009;**34**: 6902-6916. DOI: 10.1016/j.ijhydene.2009.06.013

[16] Brett DJL, Manage M, Agante E, Brandon NP, Brightman E, Brown RJC, et al. Fuels and fuel processing for low temperature fuel cells. In: Hartnig C, Roth C, editors. Polymer Electrolyte Membrane and Direct Methanol Fuel Cell Technology. Volume 1: Fundamentals and Performance of Low Temperature Fuel Cells. Sawston, Cambridge: Woodhead Publishing Limited; 2012

[17] Vielstich W. Ideal and effective efficiencies of cell reactions and comparison to Carnot cycles. In: Handbook of Fuel Cells–Fundamentals, Technology and Applications. Hoboken, New Jersey, U.S: John Wiley & Sons; 2010

[18] Lu G, Wang CY. Two-phase microfluidics, heat and mass transport in direct methanol fuel cells. In: Transport Phenomena in Fuel Cells. Billerica, Massachussetts, USA: WIT Press; 2005. pp. 317-358. DOI: 10.2495/1-85312-840-6/09

[19] Glüsen A, Müller M, Stolten D. 45% cell efficiency in DMFCs via process engineering. Fuel Cells. 2020;**20**:507-514. DOI: 10.1002/fuce.201900234

[20] Joghee P, Malik JN, Pylypenko S, O'Hayre R. A review on direct methanol fuel cells–in the perspective of energy and sustainability. MRS Energy & Sustainability. 2015;**2**:3. DOI: 10.1557/mre.2015.4

[21] Hamnett A. Introduction to fuel-cell types. In: John Wiley & Sons, editor. Handbook of Fuel Cells–Fundamentals, Technology and Applications. Hoboken, New Jersey, U.S.: John Wiley & Sons; 2010

[22] Lan R, Tao S. Ammonia as a suitable fuel for fuel cells. Frontiers in Energy Research. 2014;**2**:1-4. DOI: 10.3389/fenrg.2014.00035

[23] Wan CH, Lin CH. A composite anode with reactive methanol filter for direct methanol fuel cell. Journal of Power Sources. 2009;**186**:229-237. DOI: 10.1016/j.jpowsour.2008.10.116

[24] Nakagawa N, Sekimoto K, Masdar MS, Noda R. Reaction analysis of a direct methanol fuel cell employing a porous carbon plate operated at high methanol concentrations. Journal of Power Sources. 2009;**186**:45-51. DOI: 10.1016/j.jpowsour.2008.09.117

[25] Han J, Liu H. Real time measurements of methanol crossover in a DMFC. Journal of Power Sources. 2007;**164**:166-173. DOI: 10.1016/j.jpowsour.2006.09.105

[26] Xu C, Faghri A, Li X, Ward T. Methanol and water crossover in a passive liquid-feed direct methanol fuel cell. International Journal of Hydrogen Energy. 2010;**35**:1769-1777. DOI: 10.1016/j.ijhydene.2009.12.055

[27] Scott K. Mass transfer in flow field. In: Vielstich W, Lamm A, Gasteiger HA, editors. Handbook of Fuel Cells–Fundamentals, Technology and

Applications. Hoboken, New Jersey, U.S: John Wiley & Sons; 2010

[28] Aric AS, Baglio V, Antonucci V. Direct methanol fuel cells: History, status and perspectives. In: Electrocatalysis of Direct Methanol Fuel Cells. Weinheim, Germany: Wiley-VCH Verlag GmbH & Co. KGaA; 2009. pp. 1-78. DOI: 10.1002/9783527627707.ch1

[29] Ramli ZAC, Kamarudin SK. Platinum-based catalysts on various carbon supports and conducting polymers for direct methanol fuel cell applications: A review. Nanoscale Research Letters. 2018;**13**:410. DOI: 10.1186/s11671-018-2799-4

[30] Qin C, Tian S, Wang W, Jiang ZJ, Jiang Z, Jovanovic P, et al. Advances in platinum-based and platinum-free oxygen reduction reaction catalysts for cathodes in direct methanol fuel cells. Frontiers in Chemistry. 2022;**10**:851-859. DOI: 10.3389/fchem.2022.1073566

[31] Arikan T, Kannan AM, Kadirgan F. Binary Pt–Pd and ternary Pt–Pd–Ru nanoelectrocatalysts for direct methanol fuel cells. International Journal of Hydrogen Energy. 2013;**38**:2900-2907. DOI: 10.1016/j.ijhydene.2012.12.052

[32] Feng L, Li K, Chang J, Liu C, Xing W. Nanostructured PtRu/C catalyst promoted by CoP as an efficient and robust anode catalyst in direct methanol fuel cells. Nano Energy. 2015;**15**:462-469. DOI: 10.1016/j.nanoen.2015.05.007

[33] Khotseng L, Bangisa A, Modibedi RM, Linkov V. Electrochemical evaluation of Pt-based binary catalysts on various supports for the direct methanol fuel cell. Electrocatalysis. 2016; **7**:1-12. DOI: 10.1007/s12678-015-0282-x

[34] Long Quan D, Huu Le P. Recent advances in Pt-based binary and ternary alloy electrocatalysts for direct methanol fuel cells. In: Electrocatalysis and Electrocatalysts for a Cleaner Environment-Fundamentals and Applications. London, UK: IntechOpen; 2022. DOI: 10.5772/intechopen.95940

[35] Jayakumar A, Madheswaran DK, Kumar NM. A critical assessment on functional attributes and degradation mechanism of membrane electrode assembly components in direct methanol fuel cells. Sustainability. 2021;**13**:13938. DOI: 10.3390/su132413938

[36] Jovanovič P, Šelih VS, Šala M, Hočevar S, Ruiz-Zepeda F, Hodnik N, et al. Potentiodynamic dissolution study of PtRu/C electrocatalyst in the presence of methanol. Electrochimica Acta. 2016; **211**:851-859. DOI: 10.1016/j.electacta.2016.06.109

[37] Thompsett D. Pt alloys as oxygen reduction catalysts. In: Vielstich W, Lamm A, Gasteiger HA, editors. Handbook of Fuel Cells–Fundamentals, Technology and Applications. Hoboken, New Jersey, U.S: John Wiley & Sons; 2010

[38] Suermann M, Bensmann B, Hanke-Rauschenbach R. Degradation of proton exchange membrane (PEM) water electrolysis cells: Looking beyond the cell voltage increase. Journal of the Electrochemical Society. 2019;**166**: F645-F652. DOI: 10.1149/2.1451910jes

[39] Iulianelli A, Basile A. Sulfonated PEEK-based polymers in PEMFC and DMFC applications: A review. International Journal of Hydrogen Energy. 2012;**37**:15241-15255. DOI: 10.1016/j.ijhydene.2012.07.063

[40] Li L. Sulfonated poly (ether ether ketone) membranes for direct methanol fuel cell. Journal of Membrane Science. 2003;**226**:159-167. DOI: 10.1016/j.memsci.2003.08.018

[41] Yuda A, Ashok A, Kumar A. A comprehensive and critical review on recent progress in anode catalyst for methanol oxidation reaction. Catalysis Reviews. 2022;**64**:126-228. DOI: 10.1080/01614940.2020.1802811

[42] Zhang Z, Liu J, Wang J, Wang Q, Wang Y, Wang K, et al. Single-atom catalyst for high-performance methanol oxidation. Nature Communications. 2021;**12**:1-9. DOI: 10.1038/s41467-021-25562-y

[43] Shrivastava NK, Thombre SB, Chadge RB. Liquid feed passive direct methanol fuel cell: Challenges and recent advances. Ionics. 2016;**22**:1-23. DOI: 10.1007/s11581-015-1589-6

[44] Sgroi M, Zedde F, Barbera O, Stassi A, Sebastián D, Lufrano F, et al. Cost analysis of direct methanol fuel cell stacks for mass production. Energies. 2016;**9**:1008. DOI: 10.3390/en9121008

[45] Willert A, Tabary FZ, Zubkova T, Santangelo PE, Romagnoli M, Baumann RR. Multilayer additive manufacturing of catalyst-coated membranes for polymer electrolyte membrane fuel cells by inkjet printing. International Journal of Hydrogen Energy. 2022;**47**:20973-20986. DOI: 10.1016/j.ijhydene.2022.04.197

[46] Mauger SA, Neyerlin KC, Yang-Neyerlin AC, More KL, Ulsh M. Gravure coating for roll-to-roll manufacturing of proton-exchange-membrane fuel cell catalyst layers. Journal of the Electrochemical Society. 2018;**165**: F1012-F1018. DOI: 10.1149/2.0091813jes

[47] Stähler M, Stähler A, Scheepers F, Carmo M, Stolten D. A completely slot die coated membrane electrode assembly. International Journal of Hydrogen Energy. 2019;**44**:7053-7058. DOI: 10.1016/j.ijhydene.2019.02.016

[48] Santangelo P, Cannio M, Romagnoli M. Review of catalyst-deposition techniques for PEMFC electrodes. TECNICA ITALIANA-Italian Journal of Engineering Science. 2019;**63**:65-72. DOI: 10.18280/ti-ijes.630109

[49] Toudret P, Blachot JF, Heitzmann M, Jacques PA. Impact of the cathode layer printing process on the performance of MEA integrating PGM free catalyst. Catalysts. 2021;**11**:669. DOI: 10.3390/catal11060669

[50] Roncaglia F, Mucci A, Romagnoli M, Spinelli L, di Bona A, Biagi R. Graphite-epoxy composites for bipolar plates: The gas permeability issue. International Journal of Hydrogen Energy. 2023;**48**: 20935-20944. DOI: 10.1016/j.ijhydene.2022.12.311

[51] Edlund D. Methanol Fuel Cell Systems: Advancing Towards Commercialization. Singapore: Pan Stanford Publishing Pte. Ltd; 2011. pp. 159-160

[52] SFC Energy. Direct Methanol Fuel Cell (DMFC). 2023. Available from: https://www.sfc.com/en/technology/direct-methanol/

[53] Mergel J, Janßen H, Müller M, Wilhelm J, Stolten D. Development of direct methanol fuel cell systems for material handling applications. Journal of Fuel Cell Science and Technology. 2012;**9**(3):031011. DOI: 10.1115/1.4006490

[54] Ramsden T. Direct Methanol Fuel Cell Material Handling Equipment Demonstration. Annual Progress Report, DOE Hydrogen and Fuel Cells Program; 2011. pp. 1209-1211

[55] Metzger N, Li X. Technical and economic analysis of fuel cells for forklift applications. ACS Omega. 26 May 2022;7 (22):18267-18275

[56] Muller J. Field experience with portable DMFC products. In: Vielstich W, Lamm A, Gasteiger HA, editors. Handbook of Fuel Cells–Fundamentals, Technology and Applications. John Wiley & Sons; 2010

[57] Efoy Pro. DMFC Power Generator. Hoboken, New Jersey, U.S: John Wiley & Sons; 2023. Available from: https://www.efoy-pro.com/en/efoy/efoy-pro-12000-duo/

[58] Blue World Technologies. DMFC Power Generator. 2023. Available from: https://www.blue.world/

[59] Kimiaie N, Wedlich K, Hehemann M, Lambertz R, Müller M, Korte C, et al. Results of a 20 000 h lifetime test of a 7 kW direct methanol fuel cell (DMFC) hybrid system–degradation of the DMFC stack and the energy storage. Energy & Environmental Science. 2014;7:3013-3025. DOI: 10.1039/C4EE00749B

[60] Oorja Protonics Inc. Direct Methanol Fuel Cells. 2016. Available from: https://www.methanol.org/wp-content/uploads/2016/06/Oorja-Corporate-Presentation-_4_27_16.pdf

[61] Material Handling Power. 2023. Available from: https://www.gaoncell.com/solution/fc/fc-dm1500

[62] Blue World Technologies. Market Applications. 2023. Available from: https://www.blue.world/markets/automotive

[63] Hogarth M. Prospects of the direct methanol fuel cell. In: Hoogers G, editor. Fuel Cell Technology Handbook. Boca Raton, Florida: CRC Press; 2002

[64] Zhao X, Yin M, Ma L, Liang L, Liu C, Liao J, et al. Recent advances in catalysts for direct methanol fuel cells. Energy and Environmental Science. 2011;**4**:2736-2753. DOI: 10.1039/c1ee01307f

[65] Das S, Pérez-Ramírez J, Gong J, Dewangan N, Hidajat K, Gates BC, et al. Core-shell structured catalysts for thermocatalytic, photocatalytic, and electrocatalytic conversion of CO_2. Chemical Society Reviews. 2020;**49**:2937-3004. DOI: 10.1039/c9cs00713j

[66] Direct Methanol Fuel Cell Market Size & Share Analysis-Growth Trends & Forecasts (2023–2028). 2023. Available from: https://www.mordorintelligence.com/industry-reports/global-direct-methanol-fuel-cells-market-industry

[67] Direct Methanol Fuel Cell Market Size, Share & Trends Analysis Report by Application, Regional Outlook, Competitive Strategies, and Segment Forecasts, 2019 to 2025. 2023. Available from: https://www.grandviewresearch.com/industry-analysis/direct-methanol-fuel-cell-market

[68] Methanol Institute. Renewable Methanol. 2023. Available from: https://www.methanol.org/renewable/

Chapter 4

Preparation and Characterization of Pd Nanoparticles Supported on Graphene-Based Anode Catalysts for Direct Methanol Fuel Cells

Sabejeje Akindeji Jerome, Adebare Nurudeen Adewumi, Yi Cheng Yi, Huaneng Su and Lindiwe Khotseng

Abstract

Palladium (Pd) nanoparticles supported by graphene nanomaterials were prepared and tested in this work using methanol as the fuel. The synthesized nanoparticles were used as electrocatalysts for direct methanol fuel cell. The support materials were synthesized by modified Hummer's method and subsequently doped with nitrogen using melamine. The electrocatalysts were synthesized using modified polyol method. The synthesis method of the electrocatalyst was further modified by adjusting the pH of the electrocatalyst from 12 to 13. The structural characterization of the support materials was carried out using Fourier Transform Infrared (FT-IR) spectroscopy and Brunauer-Emmett-Teller (BET) technique while that of the electrocatalysts was also done using X-ray diffraction (XRD) and high-resolution transmission electron microscopy (HR-TEM). The elemental analysis was carried out using energy dispersive spectroscopy (EDS) to validated the presence of N-doped in Nitrogen-doped graphene oxide (NGO) and NrGO support materials and the Pd loading. The electroactivity, electron kinetics and stability of the electrocatalyst towards methanol oxidation were evaluated using cyclic voltammetry (CV), electrochemical impedance spectroscopy (EIS) and chronoamperometry (CA) respectively. The results showed that the modification of electrocatalyst by increasing the pH to 13 did not improve the activity of the electrocatalyst generally since the supported Pd catalysts synthesized by modified polyol method exhibited better electroactivity towards methanol oxidation than their pH 13 counterparts.

Keywords: methanol oxidation, graphene oxide, reduced graphene oxide, nitrogen-doped graphene oxide, nitrogen-doped reduced graphene oxide, palladium catalyst, direct methanol fuel cells

1. Introduction

The need for continuous fight against climate change which is one of the major consequences of consistent use of fossil fuel has been established. This exponential increase in fossil fuel consumption leading to increase in environmental pollution caused by emission of carbon monoxide has become a major concern. It has been reported by International Energy Agency (IEA) that in 2018, the global emission of CO_2 energy related has risen beyond 33 Gt CO_2 [1] of which over 10 Gt came from coal used in power generating plant, majorly in Asia. This was as a result of increase in energy demand despite an increase in renewable sources of 7% in 2018 which resulted to 0.5% increase in CO_2 emission for every 1% gain in economic output globally [2]. The United Intergovernmental Panel on Climate Change has recommended reduction of greenhouse gas emission by 50–85% by 2050 in order to mitigate the climate change [1]. The reduction in this hazardous emissions can be achieved by using renewable energy source such as methanol [3, 4]. Hence, the need to translate from fossil fuel based energy to alternative source of energy has become imperative [5, 6]. These have drawn the attention of government and researchers on how to develop, improve and commercialize greener alternative, renewable and sustainable sources of energy [6].

The major sources of renewable energy, which are wind and solar, are characterized by irregular and instability of weather condition and time of each day [7]. Methanol has been reported to be environmentally friendly which does not suffer from these weather conditions and can be used as a good fuel when synthesized from hydrogen by electrolysis and CO_2 from atmosphere, biomass or from exhaust of industrial processes [8, 9]. Therefore, its usage can reduce the climate change. Methanol can be CO_2 free if obtained from renewable sources. Methanol is a good substitute for oil and can be easily handled [10]. Furthermore, being a liquid at ambient condition of temperature and pressure, it ties the habit with renewable energy systems where fuel cells and electrolysers play major role [2].

China has been reported as the largest methanol producer in the world with about 70 million tons [11]. In 2018, methanol consumption was about 17.4 million tons while only road transportation consumed about 126 million tons of petrol and 156 million tons of diesel fuel [12]. The percentage of methanol usage around the world is given as: China- 58%; the rest of Asia-Pacific- 16%; Europe- 13%; Latin America- 2%; North America- 10%. The rate of methanol production in China has increased exponentially despite its reduction globally. The migration to methanol economy in China strengthened their energy security, air pollution, reduced CO_2 emission and enhance the added value of their domestic economy [1].

The use of methanol as transportation fuel in China has geometrically increased its demand in all sectors reaching up to 40% of its total methanol production despite the abundant of coal in order to ameliorate energy security and air pollution. This has invariably enhanced their energy dependence and reduction in harmful emission. Methanol vehicle pilot programme was carried out in China between 2012 and 2018 and the results showed that methanol is a viable transportation fuel with no technical, economic and safety challenges. Several thousands of vehicles are now using methanol as fuel [12]. This increase in methanol vehicle has improved their economic sustainability. This has also been extended to marine vehicles as the International Maritime Organization set new limit such that all sizes of ships need to use fuel that meet the 0.5% m/m SOx emission with effect from 1st January, 2020 for ship operating outside designated emission control areas [2].

Methanol is majorly converted into neat methanol-M100, methanol and petrol blend (M5, M10, M15, M30, M50 and M85), methanol based petrol, methyl tertiary butyl ether (MTBE), dimethyl ether (DME) and biodiesel [1].

1.1 Benefits of methanol based fuels

The use of methanol either as neat or blend fuel has major advantages over the conventional uses. Tian et al. [13] reported that M20 enhance the thermal efficiency of engines and reduction of CO, CO_2 and NOx emission [13]. Wang et al. [14] noted that M15 and M25 are more accepted than petrol [14].

Electrofuel, in which methanol is an example, has been reported to play a significant role in decarbonizing aviation [2]. Goldman et al. [15] reported five electrofuel including methanol for use in aviation and discovered that they can be good substitute for the conventional kerosene base fuels but higher structural loads and lower efficiency still remain as challenge [15].

Recently, there has been paradigm shift in methanol utilization. About 85% of methanol production was reportedly use in chemical industry in 2012. This was due to the fact that methanol exhibits some unique characteristics which placed it ahead conventional fuels for internal combustion engine. This include high octane rating, high latent heat, fast burning velocity, high compression ratio and high knock resistance which increase engine efficiency [16, 17]. Methanol has been a good fuel of choice mostly in motorsport due to its clean burning and less expensive nature, high performance, safety and high energy density [2] compared to other fuels [18–20].

This chapter therefore focuses on direct methanol fuel cells using methanol as a renewable fuel and modifying graphene supported Palladium catalyst as facilitator of methanol oxidation for easy transition from the use of fossil fuel to a sustainable renewable methanol fuel as energy source for the future.

1.2 Methanol in fuel cells

Methanol is used in fuel cells either directly in direct methanol fuel cells (DMFCs) or indirectly via methanol steam reforming into hydrogen-rich gas mixture in high temperature proton exchange membrane fuel cells (PEMFCs). This discussion is limited to DMFCs in this chapter.

1.2.1 Direct methanol fuel cells

Low-temperature fuel cells using hydrogen, methanol, ethanol and other fuel is a technology which has drawn the attention of many researchers because they serve as means of generating power by direct conversion of the fuel and oxygen into water, electrochemically [21, 22]. Direct methanol fuel cell is one of these new technology.

A DMFCs is a low-temperature PEMFCs that make use of liquid methanol as fuel. This methanol, which serves as the fuel, can be produced from biomass which is non-hazardous to the environment [23]. The operating principle of a DMFCs comprises five major porous layers which include anode gas diffusion layer (AGDL), anode catalyst layer (ACL), polymer electrolyte membrane (PEM), cathode catalyst layer (CCL) and cathode gas diffusion layer (CGDL) as shown in **Figure 1**. The methanol fed into the anode diffuses through the AGDL to ACL where it is oxidized as shown in Eq. (1). During the cell operation, Eq. (1) proceeds forwards to form carbon dioxide, protons and electrons [24].

$$CH_3OH + H_2O \rightarrow CO_2 + 6H^+ + 6e^- \quad (1)$$

The reaction in the ACL occurs in three-phase boundary which include catalyst particles, carbon support and electrolyte (membrane). The electron produced at ACL

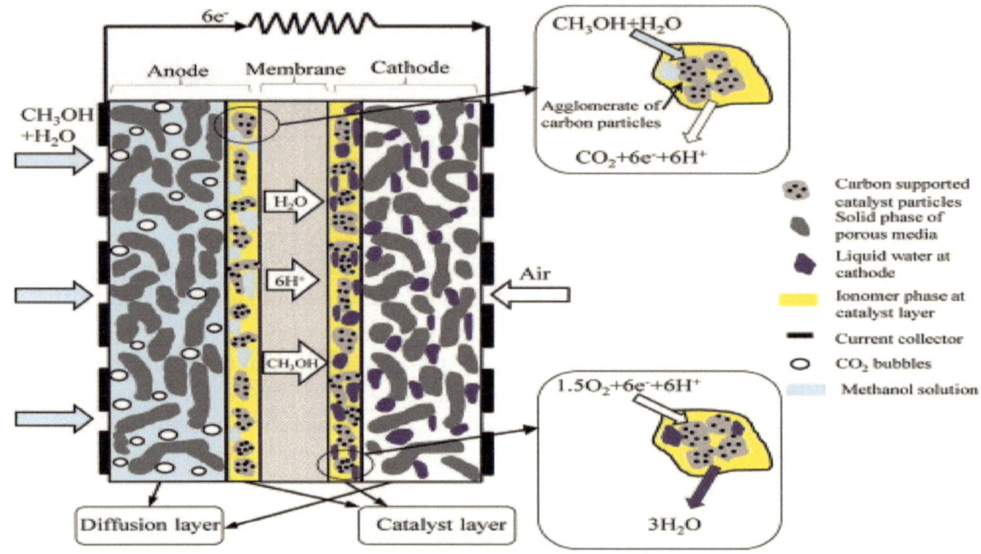

Figure 1.
Schematic of a DMFCs during normal operation. Source: [24].

are transferred through the carbon support to the AGDL where they move through the external circuit and converted to electric current while the remaining unconverted electrons move to the cathode side of the fuel cell. However, the proton generated are transferred through the ACL ionomer phase to the membrane. The membrane is impermeable to the electron and gaseous species. At the cathode, oxygen gas is being forced in as it diffuses through the CGDL to CCL where it is reduced to heat and water in the presence of electrons and protons as shown in Eq. (2) [24]. These prominent features enable DMFCs to be considered as a promising device to supply power in portable devices [5, 25, 26].

$$3O_2 + 12H^+ + 12e^- \rightarrow 6H_2O \tag{2}$$

The net equation for DMFCs reaction can be summarized as:

$$2CH_3OH + 3O_2 \rightarrow 2CO_2 + 4H_2O \tag{3}$$

The aim of DMFCs research is to develop low cost, high performance and durable cells with maximum oxidation of methanol fuel that can power portable devices [27, 28]. A lot of research has been carried out with the intension of reducing the cost and increasing the performance of fuel cells using different strategies. Some of these strategies include reducing the electrocatalyst loading in fuel cell electrodes, developing novel nanostructured thin-film Platinum such as 3 M's nanostructured thin film (NSTF) electrode, decreasing the electrocatalyst nanoparticles size, reducing Platinum dependence by developing metallic alloy either as binary or as ternary and Platinum-free electrocatalysts, improving electrocatalyst dispersion using novel fabrication methods, developing membrane electrode assembly (MEA) fabrication methods to enable better catalyst dispersion and utilization, using new techniques to increase mass transport at the fuel cells electrode surface, improving the performance of carbonaceous electrocatalyst supports and exploring novel non-carbonaceous electrocatalyst support materials [27, 28]. In contrary, the present DMFCs system is very expensive (mainly due to catalyst used) with low performance and less durability. One of the major factors to be considered in designing high performance and more durable DMFCs is the catalyst support materials [29, 30] since they have been discovered to reduce the cost, improve the catalytic activity by increasing the catalyst nanoparticle distribution and durability for maximum oxidation of methanol fuel in DMFCs if properly developed [31].

1.3 Support materials

At this juncture, it is imperative to mention that high performance of DMFCs also depend majorly on the properties of the support materials used [32]. Electrocatalyst support materials play a crucial role in enhancing electrocatalyst activity during DMFCs operation. Through their electronic and atomic structure, they provide a good surface area for homogenous dispersion, better particle size and also promote the stability of the catalyst nanoparticles [6, 28, 33–37]. Since the instability of the catalyst support materials results to detachment of catalyst nanoparticles from the support materials causing the loss of activity of the electrocatalysts, the support materials can therefore significantly influence the activity of the catalysts and prolong their stability [38, 39]. Therefore, the activity and stability of electrocatalysts are function of the type of support materials used [40] as supported metal catalysts have been discovered to show higher stability and activity compared to the unsupported ones [39].

A lot of research has been carried out on large number of carbon support materials. Due to their high availability and low cost, carbon black materials have been widely explored as support materials for Pt and Pt alloyed electrocatalysts in low-temperature fuel cells such as DMFCs [6, 21, 26, 35, 41–45]. Among the carbon black support materials developed include Vulcan XC-72, Black Pearls 2000, Acetylene Black, Ketjen Black and Mascorb and they all exhibit high surface area (>100 m^2g^{-1}) and good electrical conductivity (>1Scm^{-1}). Among these carbon black support materials, Vulcan XC-72 with BET surface area of 250 m^2g^{-1}, mesoporous and macroporous percentage of 54% and electric conductivity of 2.77 Scm^{-1} has been reported to show a significant performance in fuel cell environment [6, 26, 35, 46]. Furthermore, carbon materials with high nanoarchitectural graphitic structures such as multi-walled carbon nanotubes (MWCNTs) and carbon nanofibers (CNFs) have also been critically examined. This is as a result of their unique features as they offer better crystalline structure, high electrical conductivity, excellent corrosion resistance with high level of purity. MWCNTs in particular, is of great interest because of the specific structural, mechanical and electrical properties they exhibit [21, 38]. Mesoporous carbons (MCs) which include ordered mesoporous carbons (OMCs) have also been extensively studied as support materials for Pt and Pt alloyed electrocatalysts [6, 21]. Compared to carbon blacks, mesoporous carbon materials possess higher surface area with little or no micropores which facilitate the high dispersion of the catalyst nanoparticles on their surface and their pores. This results in large effective surface area of the electrocatalyst with high catalytic activity. Mesoporous structure with mesoporous size of 2–50 nm enhance easy mass transport, producing high limiting current value [6, 47–52].

Recently, research interest has also been diverted, towards prominent 2D graphene and its N-doped derivatives [53]. This attraction is due to their unique graphitic forms, high charger-carrier mobility (up to 105 $cm^2V^{-1}S^{-1}$), super conductivity, ambipolar electric field effect, quantum Hall effect at room temperature, high mechanical strength (130 GPa) and high surface area (2600 m^2g^{-1}) [54]. Graphene surface area contains enough oxygen functional groups which give it a better advantage over other support materials. This enables graphene to disperse any metal nanoparticles easily and efficiently. It also possesses the ability to remove a lot of accumulated carbon monoxide which act as a poison during the adsorption of the catalyst nanoparticles thereby increasing the electrocatalytic activity of the catalyst [55–59]. Moreover, N-doped graphene has also been discovered to be a good catalyst support material due to its ability to introduce chemically active sites for reaction and anchoring sites for metal nanoparticles deposition, modify electronic properties and give carbon materials a metallic character [60]. Doping of graphene with nitrogen, which serves as a strong metal-support link, facilitate reduction in CO_{ads} accumulation on the surface of the electrocatalyst, thereby increasing the catalyst poison tolerance, high electrocatalytic activity and long durability [54]. Other carbon supports that have been investigated as support materials for electrocatalysts include carbon gels (CGs), carbon nanohorns (CNHs), carbon nanocoils (CNCs), activated carbon fibers (ACFs) and boron-doped diamonds (BDDs) [61, 62].

Furthermore, since activity and stability of direct methanol fuel cells (DMFCs) are anchored on the strong chemical synergistic interaction between the catalysts and the supporting materials which determines the proper dispersion of the catalyst nanoparticles at low metal loading [6], ideal catalyst support materials should therefore contain the following features among others: sufficient electrical conductivity, large surface area, high resistance to electrochemical corrosion, suitable porosity and porous structure, strong stability in acidic or alkaline medium, good proton conductivity and crystallinity, good compatibility with electrodes, good water handling to avoid flooding and easy recovery of catalysts which all result into strong chemical synergistic interaction between the support and the catalyst nanoparticles as shown in **Figure 2** [6, 21, 42, 63]. Carbon support materials have been reported to be the best choice as catalyst support due to their large specific surface area, strong and better corrosion resistance and relatively low price [26].

Based on the review of related literature, it is highly evident that support materials are important as they contribute immensely to the activity, stability and durability of catalyst and consequently enhance the performance of the fuel cells. Despite the new carbon support materials that have been explored, the DMFCs electrocatalysts still

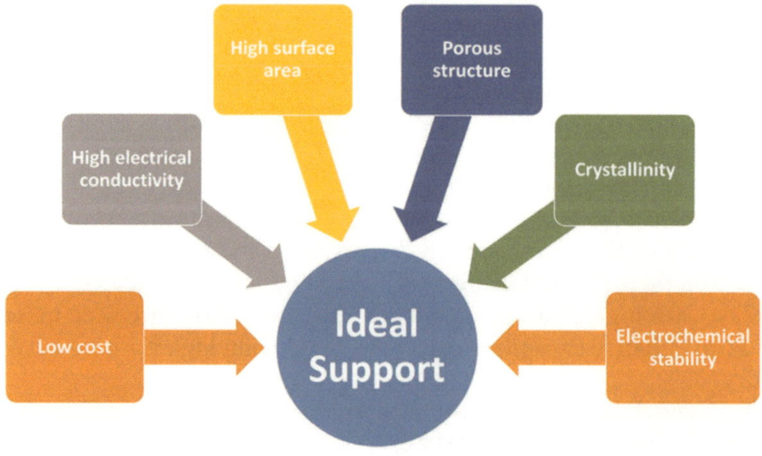

Figure 2.
Properties of an ideal catalyst support. Source: [63].

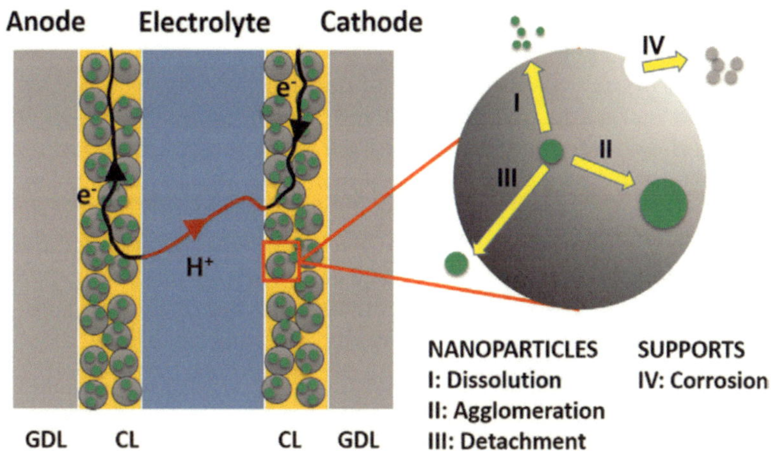

Figure 3.
Schematic illustration of electrocatalysts degradation. Source: [31].

suffer from dissolution, agglomeration, detachment from support materials and corrosion of support materials as shown in **Figure 3**. These challenges between the properties of these novel support materials and their real application under fuel cell operating system still create gaps which need to be rectified. Therefore, there is a need to modify these support materials by optimizing their properties in respect to fuel cell practical working condition by considering selection of appropriate support materials, their combination ratio (for the hybrid supports), synthesis procedure, MEA preparation and their integration into fuel cell system.

1.4 Carbon supported catalysts used in direct methanol fuel cells

Electrocatalysts has been noted to play a significant role in DMFCs architecture and have been extensively explored to enhance the rate of electrochemical reactions in order to get desirable results [64]. These catalysts are either used as anode catalyst where oxidation reaction occurs or as cathode catalyst where reduction reaction takes place. They could be developed as electrode itself or coated on the surface of the electrode. Platinum and Palladium are mostly used in DMFCs as pure metal doped on carbon support materials or as alloyed with other metals [29, 65, 66].

Platinum has been extensively used in DMFCs being the known most active metal for methanol oxidation reaction and oxygen reduction reaction among other pure metals when supported on a conductive carbon material [64, 65]. However, the activity for the methanol oxidation reaction of Pt metal alone is very low (Ermete [66]) as it suffers kinetic limitation and also readily poisoned by CO specie, a product of methanol oxidation at low temperature [67, 68]. This poisoning effect usually result to instability as well as reduction in DMFCs performance. It is also known that the corrosion of carbon black increases in the presence of Pt nanoparticles. This results to detachment of the Pt from the support and the agglomeration of the Pt nanoparticles [64]. Hence, the use of additional metal with Pt such as Ru, Ni, Co, and Mo as alloy has been developed [69, 70]. The bifunctional mechanism explains that the second metal supplies oxygen to oxidized the Pt-adsorbed methanol oxidation intermediate specie, while the electronic effect states that the second metal modifies the Pt electronic configuration, thereby weakening the adsorption of the methanol oxidation intermediate specie on Pt (Ermete [66]).

Significant efforts have also been made to develop new catalyst for DMFCs anode with little or no Pt metal and are able to tolerate poisoning by CO specie with fast kinetics [29]. In view of this, Pd has aroused notable interest as a substitute to Pt in electrocatalysts since it is more abundant in nature than Pt and exhibits the capacity to enhance the oxidation of several alcohols in alkaline media with significant electrochemical stability [29]. The attraction of Pd-based electrocatalyst emanated from the fact that, unlike Pt-based electrocatalyst, they can be highly active for oxidation of large variety of substrate in alkaline medium. The alloying of Pd with non- noble metal in catalytic architecture capable of rapidly and stably oxidizing alcohols in anode electrodes is expected to decrease the cost of the membrane electrode assembly (MEA) so as to boost the commercialization of DMFCs [29] but their performance was still found to be lower than expected [65].

Therefore, performance of different modified graphene support materials using Pd catalyst with the aim of improving the activity and stability of the electrocatalyst for maximum oxidation of methanol fuel has been investigated in this research. Palladium (Pd) is used in this study as alternative to Pt due to its lower poisoning effect, similar electronic configuration and lattice constant. It is also more abundant in nature than Pt and exhibits the capacity to enhance the oxidation of several alcohols in alkaline media with significant electrochemical stability [29].

2. Experimental section

2.1 Chemicals

All the chemicals used in this research are analytical purity grade and were used as received without any further purification. The chemicals used for the synthesis include ethanol (99%), Ethylene Glycol (99.9%), Methanol (99.9%), Nitric Acid (60%), Sulfuric Acid (90%) Potassium Hydroxide (85%), Sodium Hydroxide (98.87%) and Potassium Permanganate which were purchased from Kimix Chemical and Laboratory Suppliers, Cape Town, South Africa. Sodium Nitrate (99%), Melamine (99%), Graphite powder, Carbon nanofibers, 2-Propanol (99.5%) were purchased from Sigma- Aldrich while Hydrochloric Acid (32%) and Hydrogen Peroxide (50%) were purchased from B&M Scientific. Palladium Chloride was purchased from SA Precious Metal PTY Ltd. while Nafion solution was purchased from Ion Power Inc. The MWCNTs were bought from Carbon Nano-materials Technology Co. Ltd., Gargdong, Gyongju, Gyeonggi, South Korea with a width of ~20 nm and a length of ~10 μm. All synthesis was done using deionized water from the Milli-Q water purification system (Millipore, Bedford, MA, USA).

2.2 Synthesis of different support materials for palladium catalyst

Graphene oxide and reduced graphene oxide were synthesized using the modified Hummer's method [35, 71]. As shown in **Figure 4** 1 g of natural flake graphite powder, 0.5 g of sodium nitrate and 50 mL of sulfuric acid were mixed at 0°C ice-water bath. 3 g of potassium permanganate, being a strong oxidizing agent, was added slowly into the solution every half an hour to oxidize the graphite powder, three times in total. After that 46 mL of hot deionized water was added into the suspension drop-wise. In

Figure 4.
Synthesis procedure of graphene oxide (GO) and nitrogen-doped graphene oxide (NGO).

this step the temperature was kept at 90°C and maintained for 1 hour. Subsequently to that, 20 mL of H_2O_2 was added into the suspension drop-wise to neutralize any unreacted potassium permanganate that remains [72]. The solution was taken to an ultrasonicator for 30 minutes with the power of 200 W. The suspension was centrifuged for 30 minutes at a rotation speed 3000 rpm to remove exfoliated Graphene oxide (GO) particles [73] and a mud-like material was obtained. The material was washed with deionized water and ethanol five times, respectively. Lastly, the product was dried at 80°C in an oven for 2 days. Subsequently, the GO was reduced by dispersing 1 g of graphene oxide in 1 litre of water by means of 1 hour ultrasonic treatment as shown in **Figure 5**. As a result, a homogeneous brown graphene oxide aqueous suspension was obtained. The pH of the suspension was adjusted to 10 by addition of ammonium hydroxide while 700 μL of hydrazine solution in THF was

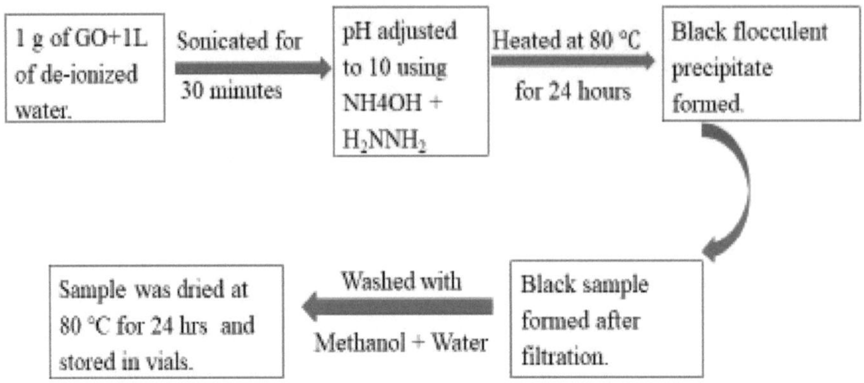

Figure 5.
Synthesis procedure of reduced graphene oxide (rGO).

added into the suspension in drops as a reducing agent [74]. The suspension was then refluxed at 80°C for a period of 24 hours. A black flocculent substance gradually precipitated out of the solution. The product was obtained by vacuum filtration process. Finally, the resulting black product was washed with methanol and ultrapure water, dried at 80°C for 24 hours in an oven and stored in vial. Thereafter, the GO, rGO and CNT were doped with nitrogen using melamine as precursor.

2.3 Synthesis of palladium catalyst using modified polyol method

About 0.4 g of the support materials were dispersed in a 15 mL of ethylene glycol under stirring conditions followed by a sonication in an ultrasonic bath for 15 minutes. To this dispersion, a solution of $PdCl_2$ in 15 mL of ethylene glycol was added and left under stirring for 15 minutes. The pH of the solution was adjusted with freshly prepared 2 M NaOH in ethylene glycol solution to pH $\sim$ 12 in the modified polyol method used for synthesizing monosupported, hybrid supported and binary catalysts. This was later modified by changing the pH from 12 to 13. The mixture was sonicated for 15 minutes to aid homogeneous adsorption of the metal precursor onto the surface of the support. For the reduction of Pd ions, the mixture was transferred into an oil bath and heated at 165°C for 6 hours consecutively under stirring and reflux conditions. After completing the reduction, the mixture was left under stirring overnight to cool down to room temperature and then filtered and washed with water. Finally, the catalyst was dried in an oven at 80°C for 24 hours in order to remove all water content and stored in vial [74–76].

3. Results and discussions

This section presents the results obtained from the various characterization carried out on all the prepared graphene support materials and their electrocatalysts using different appropriate techniques. In this study, the pH was adjusted from 12 to 13 and the results were compared in order to know if pH 13 electrocatalysts will perform better than pH 12 electrocatalysts. First, the energy dispersive X-ray spectroscopy (EDS) coupled with the scanning electron microscopy (JOEL JSM-7500F Scanning

Electron Microscope, Mundelein, II, USA) was used to evaluate the Pd metal loading and was found to be 37.67% which was the same for all the synthesized graphene supported Pd catalysts.

3.1 Surface characterization

Fourier Transform Infrared (FT-IR) and Brunauer-Emmett-Teller (BET) were used for the surface characterization of the synthesized carbon support materials while X-ray diffraction microscopy (XRD) and high-resolution transmission electron spectroscopy (HR-TEM) were used for the electrocatalysts.

3.1.1 Fourier transform infrared spectroscopy of graphene support materials

In this section, the presence of carbonyl group, hydroxyl group, nitrogen for the N-doped support materials and other functional groups in all the synthesized graphene support materials were confirmed using Fourier Transform Infrared (FT-IR) Spectroscopy. The obtained spectra are shown in **Figure 6**. For FT-IR analysis of GO, the band around 1708 and 1049 cm^{-1} were assigned to C=O and C—O of carboxylic acid respectively while the band around 1228, 1582, 2988 and 1394 cm^{-1} were assigned to C—O alcohol, C=C aromatic, C—H alkane and C—H alkane (bend) respectively. The appearance of C=O and C—O peaks of carboxylic acid in GO is an indication of the formation of GO from graphite powder by chemical oxidation [77]. In rGO spectra, the band around 1712, 3436 and 1196 cm^{-1} were assigned to C=O carboxylic acid, strong peak of O—H and C—O of alcohol respectively while the band around 1564 cm^{-1} was assigned to the C=C aromatic. The significant reduction in C=O peak of carboxylic acid, disappearance of prominent C—O peak of carboxylic acid and appearance of O—H peak of alcohol in addition to the C—O peak of alcohol in rGO show the reduction of GO to rGO [78, 79]. Furthermore, the FT-IR spectra of

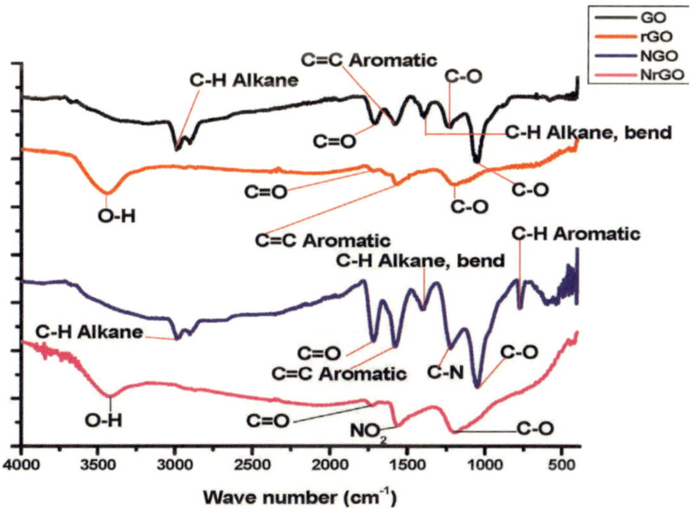

Figure 6.
The FT-IR spectra of synthesized GO, rGO, NGO and NrGO support materials.

NGO shows the C=O and C—O bands of carboxylic acid with C—N band of amine which were observed around 1716, 1050 and 1223 cm^{-1} respectively while the band around 2988, 1580, 1394 and 780 cm^{-1} were assigned to C—H alkane, C=C aromatic, C—H alkane (bend) and C—H aromatic. The appearance of C—N peak of amine which displaced the C—O of alcohol in NGO spectra is an indication of the formation of NGO from GO. In addition, the spectra of NrGO shows a medium peak of O—H band around 3414 cm^{-1} which correspond to that of hydrogen bonded alcohol; 1728, 1564 and 1188 cm^{-1} which correspond to C=O carboxylic acid, NO_2 nitro compound and C—O alcohol respectively. The significant reduction in C=O peak of carboxylic acid, disappearance of prominent C—O peak of carboxylic acid and appearance O—H peak of alcohol in addition to the C—O peak of alcohol in NrGO show the reduction of NGO to NrGO [80] while the appearance of NO_2 peak still indicate the doping with nitrogen. All these observed bands are summarized in **Table 1**. The presence of N-doped in NGO and NrGO was validated using EDS as shown in **Figure 7**.

3.1.2 Brunauer-Emmett-Teller of graphene support materials

The specific surface area, pore volume and pore size of the prepared graphene support materials were also investigated using Brunauer-Emmett-Teller (BET) as

Support Materials	Functional Groups	Observed bands (cm^{-1})
GO	C=O Carboxylic acid	1708
	C—O Carboxylic acid	1049
	C—O Alcohol	1228
	C=C Aromatic	1580
	C—H Alkane	2988
	C—H Alkane (bend)	1394
rGO	C=O Carboxylic acid	1712
	C—O Alcohol	1196
	O—H Alcohol	3436
	C=C Aromatic	1564
NGO	C=O Carboxylic acid	1716
	C—O Carboxylic acid	1050
	C—N Amine	1223
	C=C Aromatic	1580
	C—H Alkane	2988
	C—H Aromatic	780
NrGO	C=O Carboxylic acid	1728
	C—O Alcohol	1188
	NO_2 Nitro compound	1564
	O—H Hydrogen bonded alcohol	3414

Table 1.
Observed FT-IR spectra for synthesized graphene support materials.

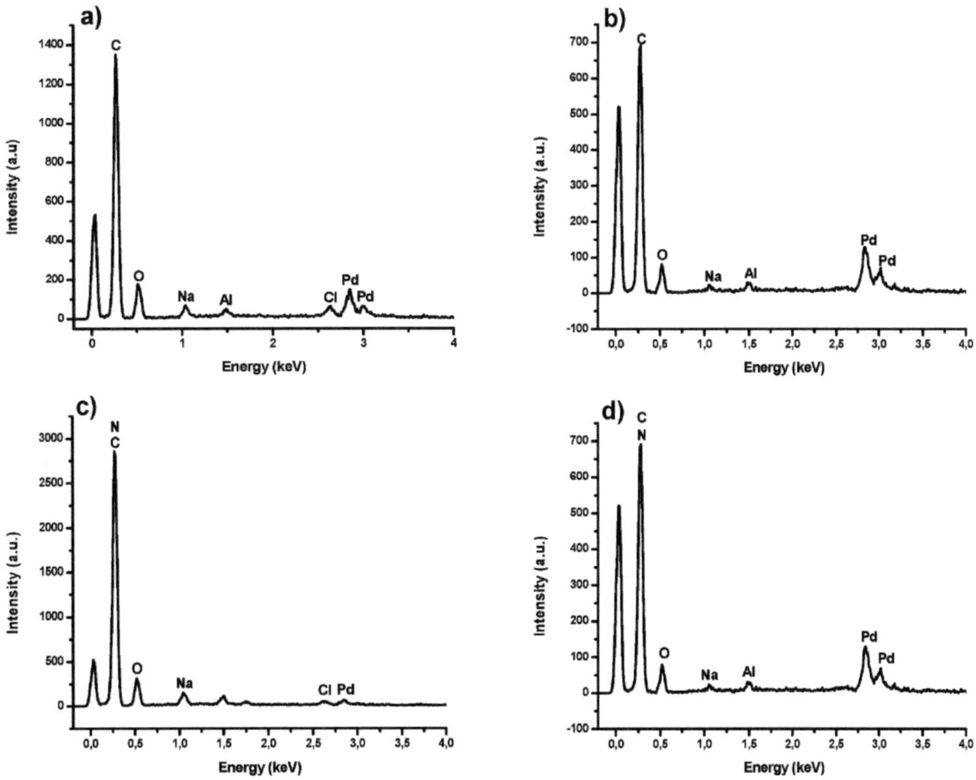

Figure 7.
The EDS spectra of synthesized (a) Pd/GO (b) Pd/rGO, (c) Pd/NGO, (d) Pd/NrGO.

presented in **Table 2**. Surface area measurements were taken from the support materials to first evaluate the surface area of the carbon support materials used. Among the prepared graphene support materials, NGO showed the highest surface area, pore volume and pore size of 41.92 $m^2\ g^{-1}$, 0.05 cm^3/g and 308.50 Å respectively. Since the performance of catalysts increases with increase in the support surface area and pore volume, the catalyst must therefore be supported with a high surface area and pore volume support materials for proper dispersion of the catalyst nanoparticles which aids the catalyst activity and make low catalyst loading feasible for fuel cell operations [21]. **Figures 8** and **9** show the adsorption–desorption and pore distribution graphs of synthesized graphene support materials respectively.

Support Materials	Surface Area (m^2/g)	Pore Volume (cm^3/g)	Pore size (Å)
GO	9.20	0.03	67.50
rGO	3.36	0.02	277.23
NGO	41.92	0.05	308.50
NrGO	6.46	0.03	173.76

Table 2.
The BET surface area, pore volume and pore size of synthesized graphene-based support materials.

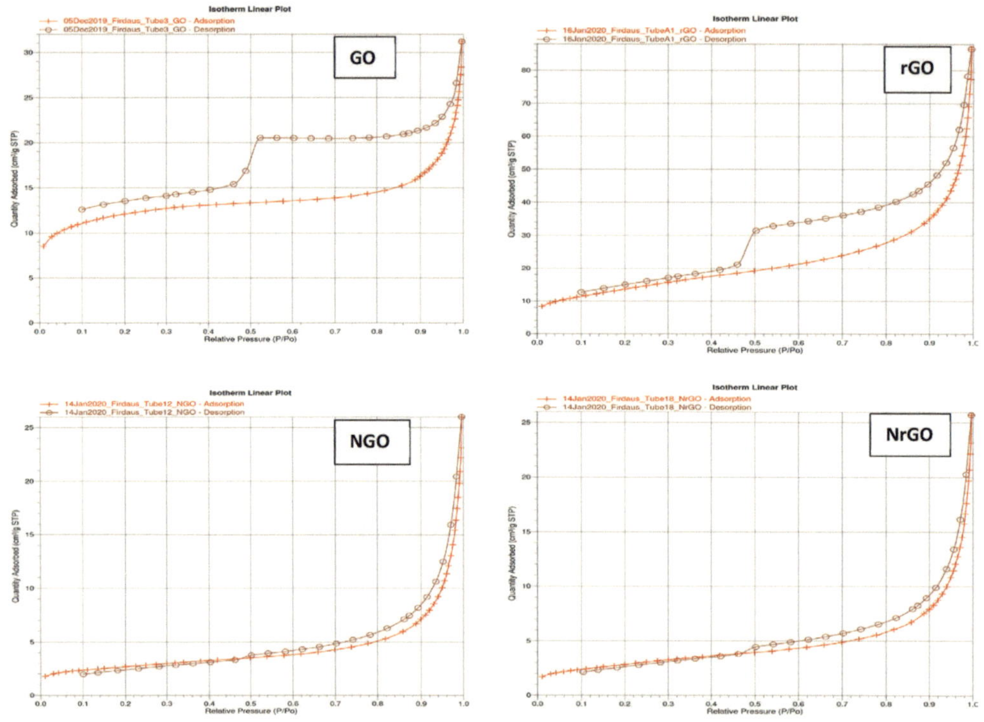

Figure 8.
Adsorption–desorption graphs of synthesized GO, rGO, NGO and NrGO.

3.1.3 X-ray diffraction of graphene supported palladium catalysts

The crystallinity and crystallite size of graphene supported Pd catalysts were determined using XRD spectra and classical Debye-Scherrer equation respectively as stated in Eq. (4).

$$d = {}^{k\lambda}/_{\beta Cos\Theta} \tag{4}$$

where d is the crystallite size, K is the Scherrer constant which also depends on the crystal shape and the diffraction line indexes, λ is the X-ray wavelength which is equal to 0.154 nm, β (2Θ) in radian is the width of the peak (full width at half maximum, (FWHM) or integral breadth) after correcting for instrumental peak broadening and Θ is the Bragg angle [81].

The sharpest and the most intense peak of all the prepared Pd catalysts appeared around 40° 2-theta scale which is indexed as (111). This peak was used to determine the crystallite size of all the electrocatalysts. The XRD graphitic pattern of prepared graphene supported Pd catalysts synthesized by modified polyol method and the modified counterparts at pH 13 show five diffraction peaks at 2-theta value around 40.0276°, 46.5107°, 68.0866°, 81.9789° and 86.8841° and are indexed to the (111), (200), (220), (311) and (222) crystal plane of Pd face-centred cubic (fcc) crystallographic structure as shown in **Figures 10** and **11** respectively [60, 82, 83]. The broad peak located at approximately 25° 2-theta scale on the other hand corresponds to the plane (002) of carbon [60, 84, 85]. The XRD spectra confirmed that all the graphene

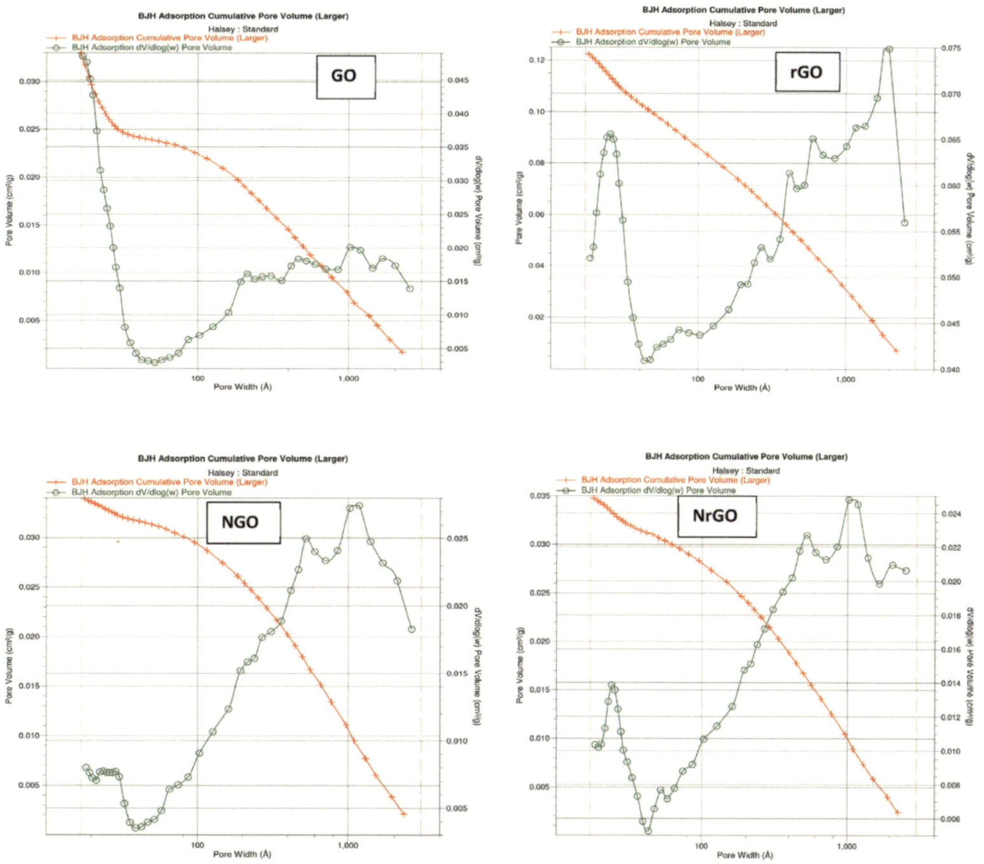

Figure 9.
Pore distribution graphs of synthesized GO, rGO, NGO and NrGO.

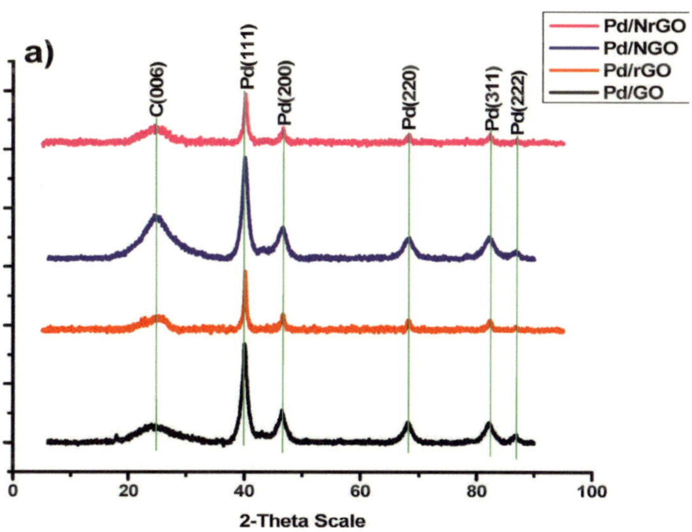

Figure 10.
XRD spectra of graphene supported Pd catalysts synthesized by modified polyol method.

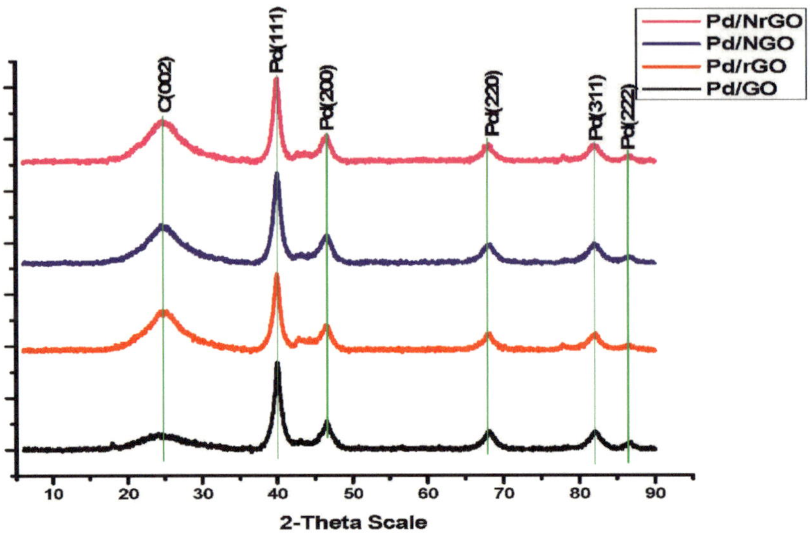

Figure 11.
XRD spectra of modified counterparts of graphene supported Pd catalyst.

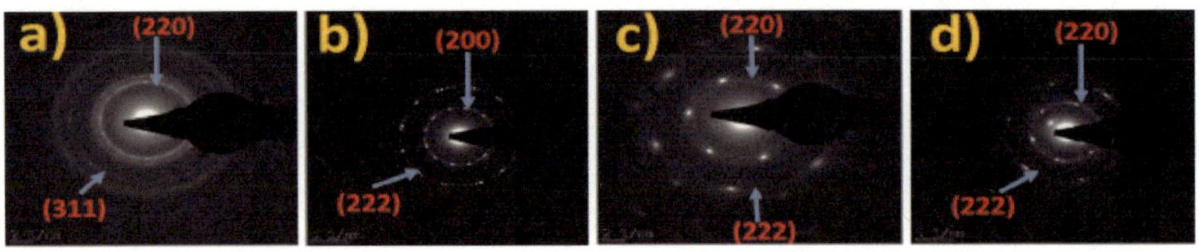

Figure 12.
Selected area electron diffraction (SAED) of graphene supported Pd catalysts synthesized by modified polyol method: (a) Pd/GO (b) Pd/rGO (c) Pd/NGO (d) Pd/NrGO.

Electrocatalyst	Particle size (nm) HR-TEM	Crystallite size (nm) XRD
Pd/GO	5 ± 1.6	5.5
Pd/rGO	19 ± 1.0	19.0
Pd/NGO	5 ± 1.2	5.8
Pd/NrGO	12 ± 1.0	12.8

Table 3.
The particle size and crystallite size of the graphene support Pd catalysts synthesized by modified polyol method.

supported Pd catalysts are crystalline as illustrated in **Figure 10**. This is also corroborated with selected area electron diffraction (SAED) as shown in **Figure 12**. The better the crystallinity, the lower the ohmic resistance and the better the electron flow [86]. The summary of particle size and crystallite size of the graphene support Pd catalysts synthesized by modified polyol method is illustrated in **Table 3**.

In case of the modified counterparts of graphene supported Pd catalysts at pH 13, the XRD spectra also revealed their crystalline structures too with the

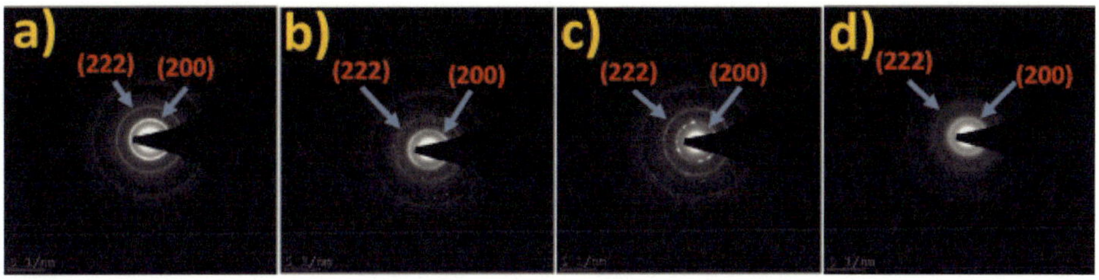

Figure 13.
Selected area electron diffraction (SAED) of modified counterparts of graphene supported Pd catalysts: (a) Pd/GO, (b) Pd/rGO, (c) Pd/NGO, (d) Pd/NrGO.

Electrocatalyst	Particle size (nm) HR-TEM	Crystallite size (nm) XRD
Pd/GO	5 ± 1.0	6.0
Pd/rGO	6 ± 1.2	6.2
Pd/NGO	5 ± 0.7	5.9
Pd/NrGO	5 ± 0.6	5.8

Table 4.
The particle size and crystallite size of the modified counterparts of graphene supported Pd catalysts.

sharpest and most intense peak indexes as (111) at 40° 2-theta scale as shown in **Figure 11**. The selected area electron diffraction (SAED) indicated in **Figure 13** also corroborate the crystallinity of all the modified electrocatalysts. The particle size and crystallite size of the modified counterparts of graphene supported Pd catalysts is illustrated in **Table 4**.

3.1.4 The high-resolution transmission electron spectroscopy of graphene supported palladium catalysts synthesized by modified polyol method

Figures 14 and **15** show the nanomorphological structures of all the synthesized graphene supported Pd catalysts examined using HR-TEM with their frequency distribution from 50 randomly selected nanoparticles. For graphene (GO, rGO, NGO and NrGO) supported Pd catalysts synthesized by modified polyol method, the images revealed a homogenous with relatively small particle size (since nanoparticles usually show a nanodimensional size of 1–100 nm [87]) distribution which ranges between 5 and 19 nm as shown in their respective histograms in **Figure 14**.

In case of modified counterparts of graphene (GO, rGO, NGO and NrGO) supported Pd catalyst, the images also revealed a homogenous distribution with relatively small particle size distribution which ranges between 5 and 6 nm as shown in their respective histograms in **Figure 15**.

3.2 Electrochemical evaluation

3.2.1 Cyclic voltammetry

The electrochemical properties of graphene supported Palladium electrocatalysts synthesized by modified polyol method and their modified counterparts in alkaline

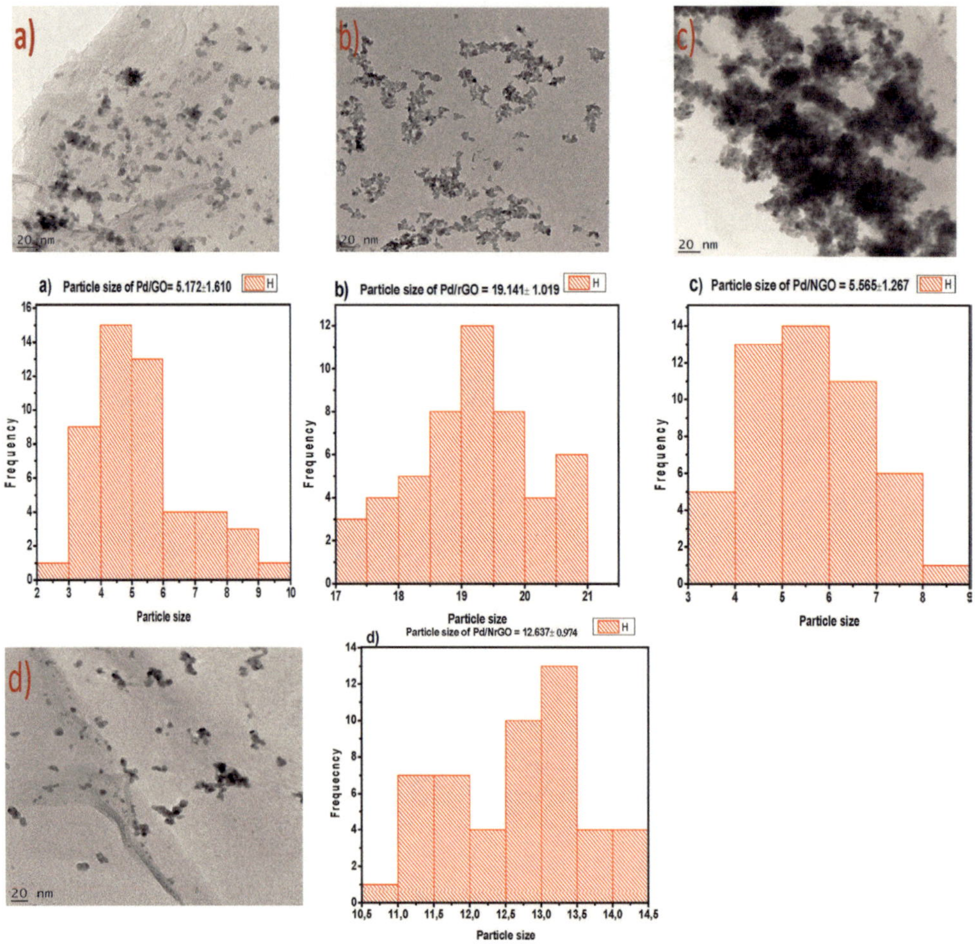

Figure 14.
HR-TEM images with their respective histograms for graphene supported Pd catalysts synthesized by modified polyol method: (a) Pd/GO (b) Pd/rGO (c) Pd/NGO and (d) Pd/NrGO.

(1 M KOH) solution were first examined by cyclic voltammetry (CV) with Pd loading of 0.02 $mgcm^{-2}$. The CV curves of each catalyst were obtained from the stabilized curve after scanning 20 cycles [88]. The cyclic voltammetry shows the adsorption/desorption peaks in the hydrogen region at negative potentials. As more negative potentials were applied, the reduction of H^+ and the adsorption of H atoms become stronger:

$$H^+_{(aq)} + e^- + \text{site} \rightarrow H_{(ad)} \quad (5)$$

This process continued as electrode potentials became more negative until the formation of a H (ad) monolayer was achieved. Immediately the Pd surface was fully covered by hydrogen atoms, the adsorption of H_2 molecules occurred:

$$2H_{(ad)} \rightarrow H_{2(ad)} \quad (6)$$

These adsorbed hydrogen molecules came together to form hydrogen bubbles which left the Pd electrode surface when they have grown large enough:

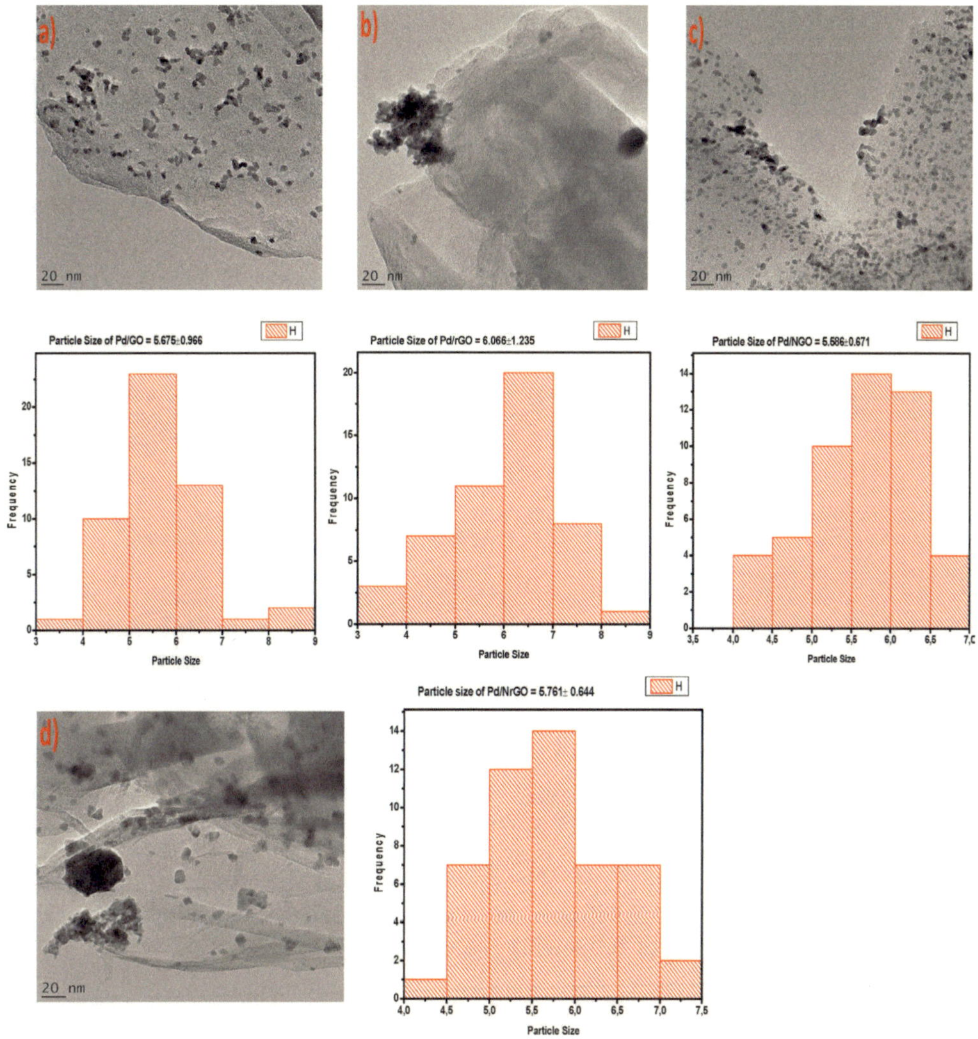

Figure 15.
HR-TEM images with their respective histograms for modified counterparts of graphene supported Pd catalysts: (a) Pd/GO (b) Pd/rGO, (c) Pd/NGO and (d) Pd/NrGO.

$$nH_{2(ad)} \rightarrow nH_{2(g)} + 2n - \text{sites} \tag{7}$$

At this period, a high cathodic potential was applied on the electrode and many free sites were exposed to the solution. Immediately the above reaction occurred at a high rate, the sharp cathodic current, known as the hydrogen evolution, increased. The formation of the H(ad) monolayer can be easily detected at the potential where the cathodic current increases rapidly. When the potential is reversed, the opposite process (anodic currents in the hydrogen region) occurs [89].

The oxidation peak of all the prepared electrocatalysts was not well pronounced [85] while a significant cathodic reduction peak which is attributed to the reduction of PdO produced on the forward potential scan was observed between −0.2 and −0.4 V [83] for all the prepared electrocatalysts. Among electrocatalysts synthesized by

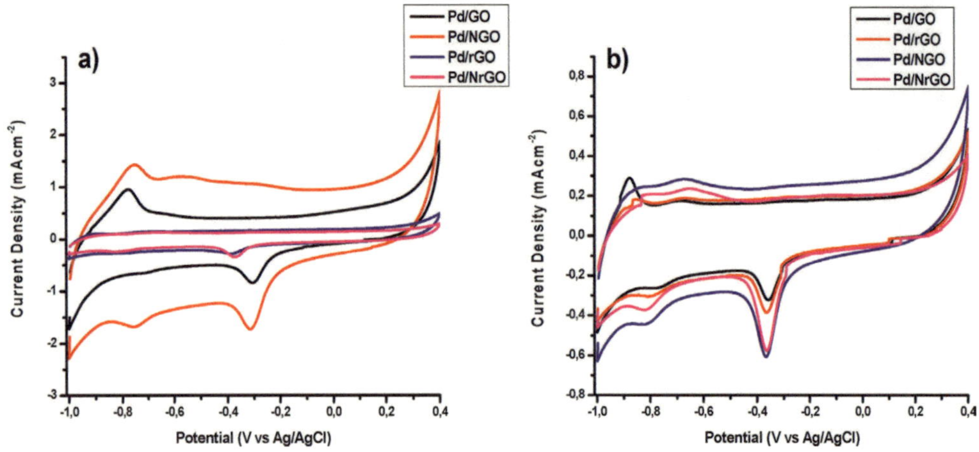

Figure 16.
The cyclic voltammetry of (a) graphene supported Pd catalysts synthesized by modified polyol method (b) modified counterparts of graphene supported Pd catalysts in N_2 saturated 1 M KOH at scan rate of 0.02 vs^{-1}.

modified polyol method, Pd/NGO exhibited the most intense cathodic reduction peak with highest current density which implies that it provided better evidence for the widest electroactive surface area (ECSA) among the graphene supported Pd catalysts synthesized by modified polyol method as shown in **Figure 16a** [90, 91]. However, for the modified electrocatalysts, Pd/NGO exhibited the most intense cathodic reduction peak as shown in **Figure 16b** which implies that it provided better evidence for the widest electroactive surface area among the modified graphene supported Pd catalysts [90, 92].

The ECSA values of all the graphene supported Pd catalysts synthesized by modified polyol method and their modified counterparts were determined by peak area of the cathodic reduction peak of PdO using the equation [83, 93]:

$$ECSA_{Pd,cat}\left(cm^2/mg\right) = Q\left(C/cm^2\right)/420\mu C/cm^2 L_{Pd}\left(mg/cm^2\right) \quad (8)$$

where Q (C/cm^2) is the charge associated with the reduction peak of the catalysts in Coulomb, L_{Pd} (mg/cm^2) is the working electrode Pd loading (0.02 mg/cm^2) while 420 $\mu C/cm^2$ is the value for oxygen monolayer of Pd in Eq. (5) [83]. From the CV results of graphene (GO, rGO, NGO and NrGO) supported electrocatalysts synthesized by modified polyol method, it is clear that Pd/NGO have the highest ECSA value

Catalysts	Electroactive Surface Area (m^2/g)	Current Density (mA/cm^2) for MOR	Current Density (mA/cm^2) for Chronoamperometry
Pd/GO	1.60	3.45	0.07
Pd/rGO	1.24	1.02	0.03
Pd/NGO	1.84	7.38	0.14
Pd/NrGO	1.53	2.99	0.11

Table 5.
Comparison of ECSA with current densities (MOR and Chrono) of graphene supported Pd catalysts synthesized by modified polyol method as determined from the anodic sweep (−0.1 to 0.4 V) at scan rate of 0.02 vs^{-1}.

Catalysts	Electroactive Surface Area (m^2/g)	Current Density (mA/cm^2) for MOR	Current Density (mA/cm^2) for Chronoamperometry
Pd/GO	1.70	2.43	0.05
Pd/rGO	3.52	2.70	0.06
Pd/NGO	3.87	3.88	0.14
Pd/NrGO	3.78	4.88	0.14

Table 6.
Comparison of ECSA with current densities (MOR and Chrono) of modified graphene supported Pd catalyst as determined from the anodic sweep (−0.1 to 0.4 V) at scan rate of 0.02 vs^{-1}.

of 1.84 m^2/g compared to other graphene supported Pd catalysts as indicated in **Table 5**. For the modified counterparts too, it is also evident from the CV results that Pd/NGO exhibited the highest ECSA value of 3.87 m^2/g among graphene supported Pd catalysts as illustrated in **Table 6**.

3.2.2 Methanol oxidation reaction

The electrocatalytic activity of the as-synthesized graphene supported Pd catalysts synthesized by modified polyol method and their modified counterparts towards methanol oxidation reaction (MOR) in alkaline (1 M KOH) solution in the presence of methanol was examined by cyclic voltammetry (CV) as illustrated in **Figure 17a**. In the forward scan, the oxidation peaks correspond to the oxidation of freshly chemosorbed species coming from methanol adsorption. The reverse scan peaks are basically associated with the removal of carbonaceous species which were not completely oxidized in the forward scan than the oxidation of freshly chemosorbed species [94]. The onset potential of graphene supported Pd catalysts synthesized by modified polyol method and their modified counterparts varies from one to another as summarized in **Tables 7** and **8** respectively. After the anodic scan, the anodic current density declined sharply as a result of the formation of PdO on the electrocatalysts surface at high anodic potential. As the backward scan commenced, the PdO began to

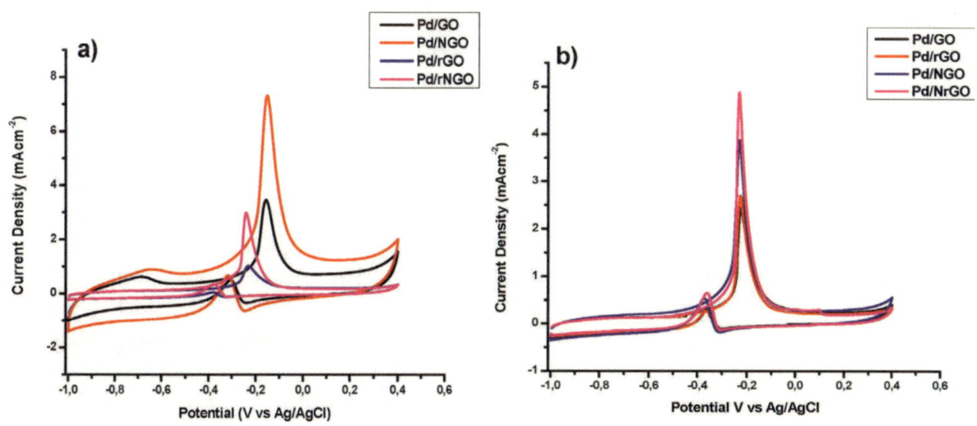

Figure 17.
The cyclic voltammetry curves of methanol oxidation on (a) graphene supported Pd catalysts synthesized by modified polyol method (b) modified graphene supported Pd catalysts in N_2 saturated 1 M MeOH +1 M KOH at scan rate of 0.02 vs^{-1}.

Electrocatalyst	Onset Potential (V vs. Ag/AgCl)	Anodic peak for forward scan I_f ($mAcm^{-2}$)	Anodic peak for reverse scan I_r ($mAcm^{-2}$)	I_f/I_r ratio
Pd/GO	−0.38	3.45	0.49	7.06
Pd/rGO	−0.42	1.02	0.05	21.79
Pd/NGO	−0.36	7.31	0.68	10.83
Pd/NrGO	−0.49	2.99	0.35	8.54

Table 7.
Results of the study of CVs of graphene supported Pd catalysts synthesized by modified polyol method in 1 M KOH + 1 M methanol (MeOH).

Electrocatalyst	Onset Potential (V vs. Ag/AgCl)	Anodic peak for forward scan I_f ($mAcm^{-2}$)	Anodic peak for reverse scan I_r ($mAcm^{-2}$)	I_f/I_r ratio
Pd/GO	−0.44	2.43	0.35	6.94
Pd/rGO	−0.45	2.70	0.25	10.76
Pd/NGO	−0.46	3.88	0.53	7.28
Pd/NrGO	−0.47	4.88	0.65	7.57

Table 8.
Results of the study of CVs of modified graphene supported Pd catalysts in 1 M KOH + 1 M MeOH.

reduce and the catalysts surface is reactivated and methanol oxidation occurred again [95]. Among the graphene (GO, rGO, NGO and NrGO) supported Pd catalysts synthesized by modified polyol method, NGO supported Pd catalyst display the highest anodic peak current density while NrGO supported Pd catalyst display the highest anodic peak current density among the modified graphene supported Pd catalysts, which implies better electroactivity towards methanol electrooxidation on forward scan of negative sweep as illustrated in **Figure 17a,b** and shown in **Tables 5** and **6** respectively. This enhanced performance of Pd/NGO as well as Pd/NrGO which also concur to their stability test, can be ascribed to their better electroactive surface area and incorporation of dopant nitrogen [22, 29, 91, 92, 96–98]. The nitrogen functional group on the surface of these support materials intensifies the electron withdrawing effect against the Pd and the decrease in electron density of Pd facilitate the oxidation of methanol fuel [90]. The N-dopant also serve as defect sites to enhance the nucleation of catalyst nanoparticles [96].

The ratio of forward anodic peak current (I_f) to reverse anodic peak current (I_r) indicate the tolerance ability of electrocatalyst to accumulation of carbonaceous products and less poisoned. This ratio is the supplementary method used to determine the CO tolerance of the catalysts. All the prepared graphene supported Pd catalysts display higher ratio values in excess of 1 which are larger than those reported in literature [95]. Large value of I_f/I_r shows higher oxidation of methanol and better CO tolerance [93, 95, 99]. From the results shown in **Tables 7** and **8**, it is observed among graphene (GO, rGO, NGO and NrGO) supported Pd catalysts synthesized by modified polyol method and their modified counterparts that Pd/rGO exhibited the highest I_f/I_r ratio of 21.79 and 10.76 respectively. Therefore, Pd/rGO show the best activity towards complete methanol oxidation in both cases. This implies that those with lower activity towards complete methanol oxidation experienced CO poisoning which practically reduces their expected performance [85].

3.2.3 Electrochemical stability

The electrochemical stability of the synthesized graphene supported Pd catalysts was also tested by chronoamperometry at −0.3 V for 30 minutes. In all the current density-time curves of the graphene supported catalysts, the oxidation current density rapidly reduced in the first 64 seconds while in their modified counterparts, it rapidly reduced in the first 20 seconds which was followed by a slower decay until it attained a steady state. The high current displayed at the beginning of stability testing could be ascribed to the double layer charging between the interface of electrode/electrolyte [85]. The gradual decrease in current density with time which was significantly observed may be attributed to poisoning of the electrocatalysts and decrease in electroactive surface area as the stability test progresses [85]. After 30 minutes' stability study in 1 M KOH + 1 M methanol solution, it was observed among graphene (GO, rGO, NGO and NrGO) supported Pd catalysts synthesized by modified polyol method that the chronoamperometric responses show a different electroactivity order to that experienced in methanol oxidation. The NrGO supported Pd catalyst show more stability than GO supported Pd catalyst while NGO and rGO supported Pd catalysts still exhibit highest and lowest stability respectively in the following order: Pd/NGO > Pd/NrGO > Pd/GO > Pd/rGO as shown in **Figure 18a**. This implies that Pd/NGO, among graphene supported Pd catalysts, still shown better stability than other synthesized electrocatalyst with current density 0.1398 $mAcm^{-2}$. This better stability in Pd/NGO, which also concur to the MOR result, can also be attributed to the better electroactive surface area and incorporation of nitrogen into the support materials. This influenced the good dispersion of Pd nanoparticles and the stability of the electrodes [90]. Also, Pd/NrGO display better stability among the modified graphene (GO, rGO, NGO and NrGO) supported Pd catalysts as shown in **Figure 18b** which also show similar MOR results [43, 63, 91, 92, 96, 98, 100, 101]. The current density for the stability test of graphene supported Pd catalysts synthesized by modified polyol method and their modified counterparts is illustrated in **Tables 5** and **6** respectively.

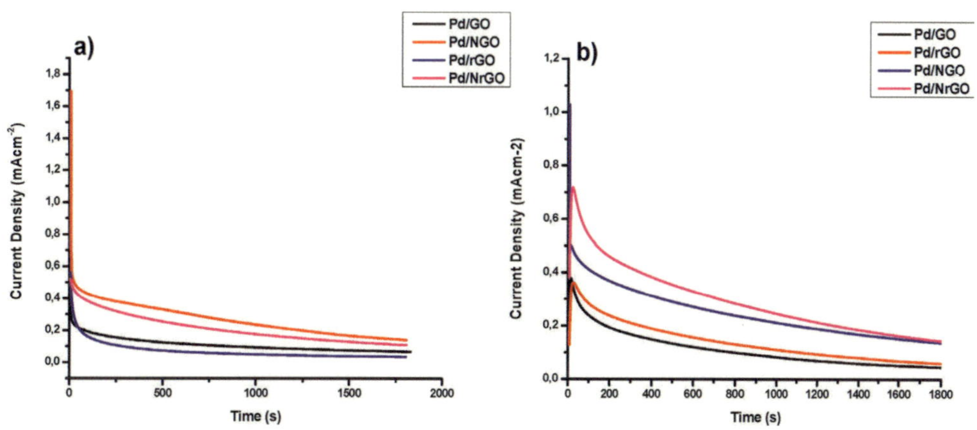

Figure 18.
The chronoamperometry of (a) graphene supported Pd catalysts synthesized by modified polyol method (b) modified graphene supported Pd catalysts in N_2 saturated 1 M MeOH +1 M KOH at potential of −0.3 V.

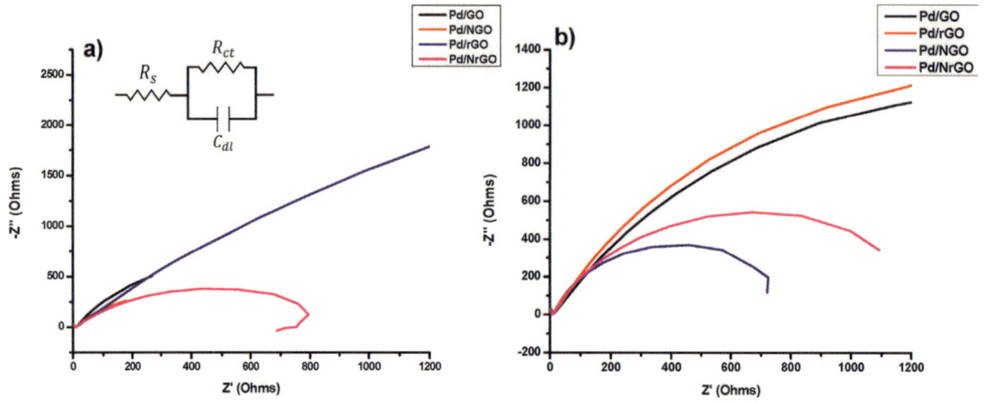

Figure 19.
The electrochemical impedance spectroscopy of (a) graphene supported Pd catalysts synthesized by modified polyol method (b) modified graphene supported Pd catalysts in N_2 saturated 1 M MeOH +1 M KOH at potential of −0.3 V.

3.2.4 Electrochemical impedance spectroscopy

The electrochemical impedance spectroscopy (EIS) revealed the thermodynamic properties of the as-synthesized electrocatalysts. It was used to explore the electrocatalytic kinetics regarding the methanol electrochemical oxidation. EIS is among the most effective techniques used to explore the electrochemical parameters of the electron/electrolyte interface [83, 102]. **Figure 19** show the interfacial behavior of the prepared electrocatalysts in KOH electrolyte containing methanol at potential of −0.3 V vs. Ag/AgCl. An equivalent circuit was employed for fitting the Nyquist plots (inset) which include solution resistance (Rs), charge transfer resistance (Rct) and double layer capacitance (Qdl). Basically, each plot shows a semicircle in the high frequency related to charge transfer. Among the graphene (GO, rGO, NGO and NrGO) supported Pd catalysts synthesized by modified polyol method, NrGO supported Pd catalyst exhibited the least electrochemical impedance. This implies that Pd catalysts supported by this support material show better chemical kinetics than other synthesized Pd catalysts as indicated by Nyquist plot in **Figure 19a** as the charge transfer kinetic of Pd catalyst on this support material significantly improved which encourage mass transfer. This was also confirmed by its resistance charge transfer (Rct) value of 0.723 $k\Omega cm^2$ which was determined using Randels-Sevcik cell fitting under open circuit as illustrated in **Table 9** which was also used for all other prepared electrocatalysts [91, 96, 98]. Furthermore, among modified graphene (GO, rGO, NGO and NrGO) supported Pd catalysts, NGO supported Pd catalysts showed the least

Electrocatalyst	Rct ($k\Omega cm^2$)	Rs ($k\Omega cm^2$)	CPE [Yo] (mF)	N (CPE Exponent)
Pd/GO	2.40	0.02	0.74	1.00
Pd/rGO	13.60	0.01	0.12	1.00
Pd/NGO	0.88	0.01	1.82	1.00
Pd/NrGO	0.72	0.06	0.22	1.00

Table 9.
Summary of electrochemical impedance spectroscopy of graphene supported Pd catalysts synthesized by modified polyol method.

DOI: http://dx.doi.org/10.5772/intechopen.1005441

Electrocatalyst	Rct (kΩcm^2)	Rs (kΩcm^2)	CPE [Yo] (mF)	N (CPE Exponent)
Pd/GO	3.37	0.03	0.47	1.00
Pd/rGO	3.43	0.03	0.46	1.00
Pd/NGO	0.71	0.03	0.61	1.00
Pd/NrGO	1.17	0.04	0.30	1.00

Table 10.
Summary of electrochemical impedance spectroscopy of modified graphene supported Pd catalysts.

electrochemical impedance. This also implies that Pd catalysts supported by this material show better chemical kinetics among their counterparts as indicated by Nyquist plot in **Figure 19b** and confirmed by its resistance charge transfer (Rct) value of 0.708 kΩcm^2 as reported in **Table 10** [29, 92, 97].

4. Conclusion

Conclusively, graphene supported Palladium catalysts synthesized by modified polyol method (Pd/GO, Pd/rGO, Pd/NGO and Pd/NrGO) were compared with modified catalysts. (Pd/GO, Pd/rGO, Pd/NGO and Pd/NrGO), it was noted that the activity of the modified ones towards methanol oxidation did not improve. The graphene supported Palladium catalysts synthesized by modified polyol method at pH 12 showed better activity towards methanol oxidation and more stability than their modified ones synthesized at pH 13 in which Pd/NGO synthesized by modified polyol method at pH 12 was identified as the best. This better performance in graphene supported Palladium catalysts synthesized by modified polyol method than their modified ones may be attributed to better dispersion of catalyst nanoparticles on their support materials and lower pH.

This study reported on graphene supported Palladium catalysts synthesized by modified polyol method (Pd/GO, Pd/rGO, Pd/NGO and Pd/NrGO) at pH values of 12 and 13. The FT-IR results showed the presence of nitrogen for the N-doped, carbonyl and hydroxyl groups on all the graphene support materials while the BET results showed the surface area of 9.20, 3.36, 41.92 and 6.46 m^2/g for GO, rGO, NGO and NrGO respectively. The XRD confirmed the crystallinity of the Palladium catalyst with average particles sizes of 5.5, 19.0, 5.8 and 12.8 nm for Pd/GO, Pd/rGO, Pd/NGO and Pd/NrGO for catalysts synthesized at pH 12. The HR-TEM results revealed that the Palladium nanoparticles were evenly distributed with little agglomerations. The EDS confirmed the presence of Palladium in the catalysts with the metal loading of 37.67%. The electrochemical characterization of the catalysts was done using cyclic voltammetry (CV), electrochemical impedance spectroscopy (EIS) and chronoamperometry (CA). The CV results showed that the catalysts synthesized at pH 12 showed better activity towards methanol oxidation; the EIS and CA also revealed that catalysts synthesized at pH 12 showed better kinetics with low electrochemical impedance and better stability respectively than those synthesized at pH 13.

In conclusion, pH 12 is a suitable pH for the synthesis of graphene supported Palladium catalysts.

Acknowledgements

This research is funded by Tertiary Education Support Programme (TESP); Eskom Holdings SOC Limited, Reg. No 2002/015527/06; National Research Foundation (NRF), South Africa, grant number 120375 and HySA systems, UWC, South Africa. We sincerely appreciate the Physics Department (UWC) for HR-TEM analysis as well as iThemba Labs for XRD analysis.

Author details

Sabejeje Akindeji Jerome[1*], Adebare Nurudeen Adewumi[2], Yi Cheng Yi[3], Huaneng Su[4] and Lindiwe Khotseng[5]

1 Department of Integrated Science, Adeyemi Federal University of Education, Ondo, Nigeria

2 Faculty of Engineering and the Built Environment, Functional Materials Research Unit (FMRU), Cape Peninsula University of Technology, Cape Town, South Africa

3 Institute of Environmental Science and Engineering, School of Metallurgy and Environment, Central South University, Changsha, China

4 Institute for Energy Research, Jiangsu University, Zhenjiang, China

5 Department of Chemistry, University of the Western Cape, Cape Town, South Africa

*Address all correspondence to: sabejerrycool@gmail.com

References

[1] Bazaluk O, Havrysh V, Nitsenko V, Baležentis T, Streimikiene D, Tarkhanova EA. Assessment of green methanol production potential and related economic and environmental benefits: The case of China. Energies. 2020;**13**(12):3113. DOI: 10.3390/en13123113

[2] Simon Araya S, Liso V, Cui X, Li N, Zhu J, Sahlin SL, et al. A review of the methanol economy: The fuel cell route. Energies. 2020;**13**(3):596. DOI: 10.3390/en13030596

[3] Yuan X, Liu X, Zuo J. The development of new energy vehicles for a sustainable future: A review. Renewable and Sustainable Energy Reviews. 2015;**42**:298-305. DOI: 10.1016/J.RSER.2014.10.016

[4] Zhang X, Liang Y, Yu E, Rao R, Xie J. Review of electric vehicle policies in China: Content summary and effect analysis. Renewable and Sustainable Energy Reviews. 2017;**70**:698-714. DOI: 10.1016/J.RSER.2016.11.250

[5] Rivera Gavidia L, Sebastián D, Pastor E, Aricò A, Baglio V. Carbon-supported Pd and PdFe alloy catalysts for direct methanol fuel cell cathodes. Materials. 2017;**10**(6):580. DOI: 10.3390/ma10060580

[6] Wang Y-J, Fang B, Li H, Bi XT, Wang H. Progress in modified carbon support materials for Pt and Pt-alloy cathode catalysts in polymer electrolyte membrane fuel cells. Progress in Materials Science. 2016;**82**:445-498. DOI: 10.1016/j.pmatsci.2016.06.002

[7] Frensch SH, Olesen AC, Simon Araya S, Kær SK. Model-supported analysis of degradation phenomena of a PEM water electrolysis cell under dynamic operation. ECS Transactions. 2018;**85**(11):37-45. DOI: 10.1149/08511.0037ECST/XML

[8] Cui X, Kær SK. Thermodynamic analyses of a moderate-temperature process of carbon dioxide hydrogenation to methanol via reverse water-gas shift with in situ water removal. Industrial and Engineering Chemistry Research. 2019;**58**(24):10559-10569. DOI: 10.1021/ACS.IECR.9B01312/SUPPL_FILE/IE9B01312_SI_001.PDF

[9] Goeppert A, Czaun M, Surya Prakash GK, Olah GA. Air as the renewable carbon source of the future: An overview of CO_2 capture from the atmosphere. Energy & Environmental Science. 2012;**5**(7):7833-7853. DOI: 10.1039/C2EE21586A

[10] Alberico E, Nielsen M. Towards a methanol economy based on homogeneous catalysis: Methanol to H_2 and CO_2 to methanol. Chemical Communications. 2015;**51**(31):6714-6725. DOI: 10.1039/C4CC09471A

[11] Methanol Price. Methanol Institute. www.methanol.org. (n.d.). Available from: https://www.methanol.org/methanol-price-supply-demand/ [Accessed: May 30, 2023]

[12] Zhao K. A Brief Review of China's Methanol Vehicle Pilot and Policy. China: Ministry of Ecology and Environment of PRC, China Vehicle Environmental Management Report 2018; 2019. Available from: http://www.miit.gov.cn/n1146295/n1652858/n1652930/n3757016/c6684042/content.html

[13] Tian Z, Zhen X, Wang Y, Liu D, Li X. Comparative study on combustion and

emission characteristics of methanol, ethanol and butanol fuel in TISI engine. Fuel. 2020;**259**:116199. DOI: 10.1016/J.FUEL.2019.116199

[14] Wang X, Ge Y, Liu L, Peng Z, Hao L, Yin H, et al. Evaluation on toxic reduction and fuel economy of a gasoline direct injection- (GDI-) powered passenger car fueled with methanol–gasoline blends with various substitution ratios. Applied Energy. 2015;**157**: 134-143. DOI: 10.1016/J.APENERGY.2015.08.023

[15] Goldmann A, Sauter W, Oettinger M, Kluge T, Schröder U, Seume JR, et al. A study on electrofuels in aviation. Energies. 2018;**11**(2):392. DOI: 10.3390/EN11020392

[16] Bertau M, Offermanns H, Plass L, Schmidt F, Wernicke HJ. Methanol: The Basic Chemical and Energy Feedstock of the Future: Asinger's Vision Today. Heidelberg, New York, Dordrecht, London: Springer; 2014. pp. i-xxxi. DOI: 10.1007/978-3-642-39709-7/COVER

[17] Zhen X, Wang Y. An overview of methanol as an internal combustion engine fuel. Renewable and Sustainable Energy Reviews. 2015;**52**:477-493. DOI: 10.1016/J.RSER.2015.07.083

[18] Gong A, Verstraete D. Fuel cell propulsion in small fixed-wing unmanned aerial vehicles: Current status and research needs. International Journal of Hydrogen Energy. 2017;**42**(33): 21311-21333. DOI: 10.1016/J.IJHYDENE.2017.06.148

[19] González-Espasandín Ó, Leo TJ, Navarro-Arévalo E. Fuel cells: A real option for unmanned aerial vehicles propulsion. The Scientific World Journal. 2014;**2014**:497642. DOI: 10.1155/2014/497642

[20] Kang K, Park S, Cho SO, Choi K, Ju H. Development of lightweight 200-W direct methanol fuel cell system for unmanned aerial vehicle applications and flight demonstration. Fuel Cells. 2014;**14**(5):694-700. DOI: 10.1002/FUCE.201300244

[21] Antolini E. Carbon supports for low-temperature fuel cell catalysts. Applied Catalysis B: Environmental. 2009; **88**(1–2):1-24. DOI: 10.1016/j.apcatb.2008.09.030

[22] Calderón Gómez J, Moliner R, Lázaro M. Palladium-based catalysts as electrodes for direct methanol fuel cells: A last ten years review. Catalysts. 2016; **6**(9):130. DOI: 10.3390/catal6090130

[23] Hasegawa F, Yokoyama S, Imou K. Methanol or ethanol produced from woody biomass: Which is more advantageous? Bioresource Technology. 2010;**101**(1):S109-S111. DOI: 10.1016/j.biortech.2009.05.008

[24] Bahrami H, Faghri A. Review and advances of direct methanol fuel cells: Part II: Modeling and numerical simulation. Journal of Power Sources. 2013;**230**:303-320. DOI: 10.1016/j.jpowsour.2012.12.009

[25] Gálvez M, Calvillo L, Alegre C, Sebastián D, Suelves I, Pérez-Rodríguez S, et al. Nanostructured carbon materials as supports in the preparation of direct methanol fuel cell electrocatalysts. Catalysts. 2013;**3**(3):671-682. DOI: 10.3390/catal3030671

[26] Tang S, Sun G, Qi J, Sun S, Guo J, Xin Q, et al. Review of new carbon materials as catalyst supports in direct alcohol fuel cells. Chinese Journal of Catalysis. 2010;**31**(1):12-17. DOI: 10.1016/S1872-2067(09)60034-6

[27] Ramli ZAC, Kamarudin SK. Platinum-based catalysts on various

carbon supports and conducting polymers for direct methanol fuel cell applications: A review. Nanoscale Research Letters. 2018;**13**(1):410. DOI: 10.1186/s11671-018-2799-4

[28] Sharma S, Pollet BG. Support materials for PEMFC and DMFC electrocatalysts—A review. Journal of Power Sources. 2012;**208**:96-119. DOI: 10.1016/j.jpowsour.2012.02.011

[29] Bianchini C, Shen PK. Palladium-based electrocatalysts for alcohol oxidation in half cells and in direct alcohol fuel cells. Chemical Reviews. 2009;**109**(9):4183-4206. DOI: 10.1021/cr9000995

[30] Chalgin A, Song C, Tao P, Shang W, Deng T, Wu J. Effect of supporting materials on the electrocatalytic activity, stability and selectivity of noble metal-based catalysts for oxygen reduction and hydrogen evolution reactions. Progress in Natural Science: Materials International. 2020;**30**(3):289-297. DOI: 10.1016/j.pnsc.2020.01.003

[31] Du L, Shao Y, Sun J, Yin G, Liu J, Wang Y. Advanced catalyst supports for PEM fuel cell cathodes. Nano Energy. 2016;**29**:314-322. DOI: 10.1016/j.nanoen.2016.03.016

[32] Munjewar SS, Thombre SB, Mallick RK. A comprehensive review on recent material development of passive direct methanol fuel cell. Ionics. 2017;**23**(1):1-18. DOI: 10.1007/s11581-016-1864-1

[33] Basri S, Kamarudin SK, Daud WRW, Yaakub Z. Nanocatalyst for direct methanol fuel cell (DMFC). International Journal of Hydrogen Energy. 2010;**35**(15):7957-7970. DOI: 10.1016/j.ijhydene.2010.05.111

[34] Ju H, Lee K-S, Um S. Multi-dimensional modeling of CO poisoning effects on proton exchange membrane fuel cells (PEMFCs). Journal of Mechanical Science and Technology. 2008;**22**(5):991-998. DOI: 10.1007/s12206-008-0207-1

[35] Priya K, Rajasekar N, Santhosh C, Nirmala Grace A. Preparation and characterization of graphene supported palladium nanoparticles for direct methanol fuel cells. In: 2014 International Conference on Advances in Electrical Engineering (ICAEE), January. Vellore, India: IEEE; 2014. pp. 1-6. DOI: 10.1109/ICAEE.2014.6838553

[36] Rodríguez-reinoso F. The role of carbon materials in heterogeneous catalysis. Carbon. 1998;**36**(3):159-175. DOI: 10.1016/S0008-6223(97)00173-5

[37] Wang Y, Zang J, Dong L, Pan H, Yuan Y, Wang Y. Graphitized nanodiamond supporting PtNi alloy as stable anodic and cathodic electrocatalysts for direct methanol fuel cell. Electrochimica Acta. 2013;**113**:583-590. DOI: 10.1016/j.electacta.2013.09.091

[38] Khotseng L, Bangisa A, Modibedi RM, Linkov V. Electrochemical evaluation of Pt-based binary catalysts on various supports for the direct methanol fuel cell. Electrocatalysis. 2016;**7**(1):1-12. DOI: 10.1007/s12678-015-0282-x

[39] Li L, Hu L, Li J, Wei Z. Enhanced stability of Pt nanoparticle electrocatalysts for fuel cells. Nano Research. 2015;**8**(2):418-440. DOI: 10.1007/s12274-014-0695-5

[40] Zhao L, Wang Z-B, Li J-L, Zhang J-J, Sui X-L, Zhang L-M. Hybrid of carbon-supported Pt nanoparticles and three dimensional graphene aerogel as high stable electrocatalyst for methanol electrooxidation. Electrochimica Acta.

2016;**189**:175-183. DOI: 10.1016/j.electacta.2015.12.072

[41] Chen LH, Zang JB, Wang YH, Bian LY. Electrochemical oxidation of nitrite on nanodiamond powder electrode. Electrochimica Acta. 2008;**53**(8):3442-3445. DOI: 10.1016/j.electacta.2007.12.023

[42] Li W, Liang C, Zhou W, Qiu J, Zhou, Sun G, Xin Q. Preparation and characterization of multiwalled carbon nanotube-supported platinum for cathode catalysts of direct methanol fuel cells. The Journal of Physical Chemistry B. 2003;**107**(26):6292-6299. DOI: 10.1021/jp022505c

[43] Sahoo M, Scott K, Ramaprabhu S. Platinum decorated on partially exfoliated multiwalled carbon nanotubes as high performance cathode catalyst for PEMFC. International Journal of Hydrogen Energy. 2015;**40**(30):9435-9443. DOI: 10.1016/j.ijhydene.2015.05.113

[44] Song H, Yang L, Tang Y, Yan D, Liu C, Luo S. Three-dimensional nitrogen-doped reduced graphene oxide–carbon nanotubes architecture supporting ultrafine palladium nanoparticles for highly efficient methanol electro-oxidation. Chemistry – A European Journal. 2015;**21**(46):16631-16638. DOI: 10.1002/chem.201502804

[45] You PY, Kamarudin SK. Recent progress of carbonaceous materials in fuel cell applications: An overview. Chemical Engineering Journal. 2017;**309**:489-502. DOI: 10.1016/j.cej.2016.10.051

[46] Alexeyeva N, Sarapuu A, Tammeveski K, Vidal-Iglesias FJ, Solla-Gullón J, Feliu JM. Electroreduction of oxygen on vulcan carbon supported Pd nanoparticles and Pd–M nanoalloys in acid and alkaline solutions. Electrochimica Acta. 2011;**56**(19):6702-6708. DOI: 10.1016/j.electacta.2011.05.058

[47] Maiyalagan T, Scott K. Performance of carbon nanofiber supported Pd–Ni catalysts for electro-oxidation of ethanol in alkaline medium. Journal of Power Sources. 2010;**195**(16):5246-5251. DOI: 10.1016/j.jpowsour.2010.03.022

[48] Park C, Baker RTK. Catalytic behavior of graphite nanofiber supported nickel particles. 2. The influence of the nanofiber structure. The Journal of Physical Chemistry B. 1998;**102**(26):5168-5177. DOI: 10.1021/jp981210p

[49] Park C, Baker RTK. Catalytic behavior of graphite nanofiber supported nickel particles. 3. The effect of chemical blocking on the performance of the system. The Journal of Physical Chemistry B. 1999;**103**(13):2453-2459. DOI: 10.1021/jp983802d

[50] Park I-S, Park K-W, Choi J-H, Park CR, Sung Y-E. Electrocatalytic enhancement of methanol oxidation by graphite nanofibers with a high loading of PtRu alloy nanoparticles. Carbon. 2007;**45**(1):28-33. DOI: 10.1016/j.carbon.2006.08.011

[51] Steigerwalt ES, Deluga GA, Cliffel DE, Lukehart CM. A Pt–Ru/graphitic carbon nanofiber nanocomposite exhibiting high relative performance as a direct-methanol fuel cell anode catalyst. The Journal of Physical Chemistry B. 2001;**105**(34):8097-8101. DOI: 10.1021/jp011633i

[52] Stein A. Advances in microporous and mesoporous solids—Highlights of recent Progress. Advanced Materials. 2003;**15**(10):763-775. DOI: 10.1002/adma.200300007

[53] Farooqui UR, Ahmad AL, Hamid NA. Graphene oxide: A promising membrane material for fuel cells. Renewable and Sustainable Energy Reviews. 2018;**82**:714-733. DOI: 10.1016/j.rser.2017.09.081

[54] Seselj N, Engelbrekt C, Zhang J. Graphene-supported platinum catalysts for fuel cells. Science Bulletin. 2015; **60**(9):864-876. DOI: 10.1007/s11434-015-0745-8

[55] Antolini E. Formation of carbon-supported PtM alloys for low temperature fuel cells: A review. Materials Chemistry and Physics. 2003; **78**(3):563-573. DOI: 10.1016/S0254-0584(02)00389-9

[56] Choi SM, Seo MH, Kim HJ, Kim WB. Synthesis of surface-functionalized graphene nanosheets with high Pt-loadings and their applications to methanol electrooxidation. Carbon. 2011;**49**(3):904-909. DOI: 10.1016/j.carbon.2010.10.055

[57] Muneendra Prasad A, Santhosh C, Nirmala Grace A. Carbon nanotubes and polyaniline supported Pt nanoparticles for methanol oxidation towards DMFC applications. Applied Nanoscience. 2012; **2**(4):457-466. DOI: 10.1007/s13204-012-0061-4

[58] Nirmala Grace A, Pandian K. Pt, Pt–Pd and Pt–Pd/Ru nanoparticles entrapped polyaniline electrodes – A potent electrocatalyst towards the oxidation of glycerol. Electrochemistry Communications. 2006;**8**(8):1340-1348. DOI: 10.1016/j.elecom.2006.06.007

[59] Selvaraj V, Grace AN, Alagar M. Electrocatalytic oxidation of formic acid and formaldehyde on nanoparticle decorated single walled carbon nanotubes. Journal of Colloid and Interface Science. 2009;**333**(1):254-262. DOI: 10.1016/j.jcis.2009.01.020

[60] Dector A, Cuevas-Muñiz FM, Guerra-Balcázar M, Godínez LA, Ledesma-García J, Arriaga LG. Glycerol oxidation in a microfluidic fuel cell using Pd/C and Pd/MWCNT anodes electrodes. International Journal of Hydrogen Energy. 2013;**38**(28): 12617-12622. DOI: 10.1016/j.ijhydene.2012.12.030

[61] Bulushev DA, Yuranov I, Suvorova EI, Buffat PA, Kiwi-Minsker L. Highly dispersed gold on activated carbon fibers for low-temperature CO oxidation. Journal of Catalysis. 2004;**224**(1):8-17. DOI: 10.1016/j.jcat.2004.02.014

[62] Kasuya D, Yudasaka M, Takahashi K, Kokai F, Iijima S. Selective production of Single-Wall carbon nanohorn aggregates and their formation mechanism. The Journal of Physical Chemistry B. 2002;**106**(19):4947-4951. DOI: 10.1021/jp020387n

[63] Anwar MT, Yan X, Asghar MR, Husnain N, Shen S, Luo L, et al. Recent advances in hybrid support material for Pt-based electrocatalysts of proton exchange membrane fuel cells. International Journal of Energy Research. 2019;**43**(7):2694-2721. DOI: 10.1002/er.4322

[64] Samad S, Loh KS, Wong WY, Lee TK, Sunarso J, Chong ST, et al. Carbon and non-carbon support materials for platinum-based catalysts in fuel cells. International Journal of Hydrogen Energy. 2018;**43**(16):7823-7854. DOI: 10.1016/j.ijhydene.2018.02.154

[65] Antolini E, Lopes T, Gonzalez ER. An overview of platinum-based catalysts as methanol-resistant oxygen reduction materials for direct methanol fuel cells. Journal of Alloys and Compounds. 2008;

461(1–2):253-262. DOI: 10.1016/J.JALLCOM.2007.06.077

[66] Antolini E. Photo-assisted methanol oxidation on Pt-TiO_2 catalysts for direct methanol fuel cells: A short review. Applied Catalysis B: Environmental. 2018;**237**:491-503. DOI: 10.1016/J.APCATB.2018.06.029

[67] Gottesfeld S, Zawodzinski TA. Polymer electrolyte fuel cells. Advances in Electrochemical Science and Engineering. 2008;**5**:195-301. DOI: 10.1002/9783527616794.CH4

[68] Pollet BG, Staffell I, Shang JL. Current status of hybrid, battery and fuel cell electric vehicles: From electrochemistry to market prospects. Electrochimica Acta. 2012;**84**:235-249. DOI: 10.1016/J.ELECTACTA.2012.03.172

[69] Shukla AK, Raman RK, Choudhury NA, Priolkar KR, Sarode PR, Emura S, et al. Carbon-supported Pt–Fe alloy as a methanol-resistant oxygen-reduction catalyst for direct methanol fuel cells. Journal of Electroanalytical Chemistry. 2004;**563**(2):181-190. DOI: 10.1016/J.JELECHEM.2003.09.010

[70] Xiong L, Manthiram A. Synthesis and characterization of methanol tolerant Pt/TiOx/C nanocomposites for oxygen reduction in direct methanol fuel cells. Electrochimica Acta. 2004;**49**(24):4163-4170. DOI: 10.1016/J.ELECTACTA.2004.04.011

[71] Eigler S, Enzelberger-Heim M, Grimm S, Hofmann P, Kroener W, Geworski A, et al. Wet chemical synthesis of graphene. Advanced Materials. 2013;**25**(26):3583-3587. DOI: 10.1002/adma.201300155

[72] Zarrin H, Higgins D, Jun Y, Chen Z, Fowler M. Functionalized graphene oxide nanocomposite membrane for low humidity and high temperature proton exchange membrane fuel cells. The Journal of Physical Chemistry C. 2011;**115**(42):20774-20781. DOI: 10.1021/jp204610j

[73] Zhao Y, Zhou Y, O'Hayre R, Shao Z. Electrocatalytic oxidation of methanol on Pt catalyst supported on nitrogen-doped graphene induced by hydrazine reduction. Journal of Physics and Chemistry of Solids. 2013;**74**(11):1608-1614. DOI: 10.1016/j.jpcs.2013.06.004

[74] Nassr ABAA, Quetschke A, Koslowski E, Bron M. Electrocatalytic oxidation of formic acid on Pd/MWCNTs nanocatalysts prepared by the polyol method. Electrochimica Acta. 2013;**102**:202-211. DOI: 10.1016/j.electacta.2013.03.173

[75] Daoush WM, Imae T. Syntheses and characterizations of multiwalled carbon nanotubes-supported palladium nanocomposites. Journal of Materials Research. 2012;**27**(13):1680-1687. DOI: 10.1557/jmr.2012.123

[76] Farsadrooh M, Yazdan-Abad MZ, Noroozifar M, Javadian H, Alfi N, Modarresi-Alam AR. Fast improved polyol method for synthesis of Pd/C catalyst with high performance toward ethanol electrooxidation. International Journal of Hydrogen Energy. 2020;**45**(51):27312-27319. DOI: 10.1016/j.ijhydene.2020.07.149

[77] Li J, Liu J, Tan G, Jiang J, Peng S, Deng M, et al. High-sensitivity paracetamol sensor based on Pd/graphene oxide nanocomposite as an enhanced electrochemical sensing platform. Biosensors and Bioelectronics. 2014;**54**:468-475. DOI: 10.1016/J.BIOS.2013.11.001

[78] Al-Marri AH, Khan M, Shaik MR, Mohri N, Adil SF, Kuniyil M, et al. Green synthesis of Pd@graphene nanocomposite: Catalyst for the selective oxidation of alcohols. Arabian Journal of Chemistry. 2016;**9**(6):835-845. DOI: 10.1016/J.ARABJC.2015.12.007

[79] Yao T, Zhang J, Zuo Q, Wang H, Wu J, Zhang X, et al. A simple way to prepare reduced graphene oxide nanosheets/Fe_2O_3-Pd/N-doped carbon nanosheets and their application in catalysis. Journal of Colloid and Interface Science. 2016;**468**:62-69. DOI: 10.1016/J.JCIS.2016.01.027

[80] Kuniyil M, Shanmukha Kumar JV, Adil SF, Shaik MR, Khan M, Assal ME, et al. One-pot synthesized Pd@N-doped graphene: An efficient catalyst for Suzuki–Miyaura couplings. Catalysts. 2019;**9**(5):469. DOI: 10.3390/CATAL9050469

[81] Uvarov V, Popov I. Metrological characterization of X-ray diffraction methods for determination of crystallite size in nano-scale materials. Materials Characterization. 2007;**58**(10):883-891. DOI: 10.1016/j.matchar.2006.09.002

[82] Carrera-Cerritos R, Baglio V, Aricò AS, Ledesma-García J, Sgroi MF, Pullini D, et al. Improved Pd electro-catalysis for oxygen reduction reaction in direct methanol fuel cell by reduced graphene oxide. Applied Catalysis B: Environmental. 2014;**144**:554-560. DOI: 10.1016/j.apcatb.2013.07.057

[83] Klaas L, Modibedi M, Mathe M, Su H, Khotseng L. Electrochemical studies of Pd-based anode catalysts in alkaline medium for direct glycerol fuel cells. Catalysts. 2020;**10**(9):968. DOI: 10.3390/catal10090968

[84] Groves MN, Malardier-Jugroot C, Jugroot M. Improving platinum catalyst durability with a doped graphene support. The Journal of Physical Chemistry C. 2012;**116**(19):10548-10556. DOI: 10.1021/jp203734d

[85] Yi Q, Chu H, Chen Q, Yang Z, Liu X. High performance Pd, PdNi, PdSn and PdSnNi nanocatalysts supported on carbon nanotubes for electrooxidation of C2C4 alcohols. Electroanalysis. 2015;**27**(2):388-397. DOI: 10.1002/elan.201400423

[86] Rao V, Simonov PA, Savinova ER, Plaksin GV, Cherepanova SV, Kryukova GN, et al. The influence of carbon support porosity on the activity of PtRu/Sibunit anode catalysts for methanol oxidation. Journal of Power Sources. 2005;**145**(2):178-187. DOI: 10.1016/j.jpowsour.2004.12.064

[87] Cookson J. The preparation of palladium nanoparticles. Platinum Metals Review. 2012;**56**(2):83-98. DOI: 10.1595/147106712X632415

[88] Zhao H, Zhao TS. Highly active carbon nanotube-supported Pd electrocatalyst for oxidation of formic acid prepared by etching copper template method. International Journal of Hydrogen Energy. 2013;**38**(3):1391-1396. DOI: 10.1016/j.ijhydene.2012.11.009

[89] Doña Rodríguez JM, Melián JAH, Peña JP. Determination of the real surface area of Pt electrodes by hydrogen adsorption using cyclic voltammetry. Journal of Chemical Education. 2000;**77**(9):1195-1197. DOI: 10.1021/ED077P1195

[90] Maiyalagan T, Viswanathan B, Varadaraju UV. Nitrogen containing carbon nanotubes as supports for Pt–alternate anodes for fuel cell applications. Electrochemistry Communications. 2005;**7**(9):905-912. DOI: 10.1016/j.elecom.2005.07.007

[91] Narreddula M, Balaji R, Ramya K, Rajalakshmi N, Ramachandraiah A. Nitrogen doped graphene supported Pd as hydrogen evolution catalyst for electrochemical methanol reformation. International Journal of Hydrogen Energy. 2019;**44**(10):4582-4591. DOI: 10.1016/j.ijhydene.2019.01.037

[92] Bock C, Paquet C, Couillard M, Botton GA, MacDougall BR. Size-selected synthesis of PtRu nano-catalysts: Reaction and size control mechanism. Journal of the American Chemical Society. 2004;**126**(25): 8028-8037. DOI: 10.1021/ja0495819

[93] Garsany Y, Baturina OA, Swider-Lyons KE, Kocha SS. Experimental methods for quantifying the activity of platinum electrocatalysts for the oxygen reduction reaction. Analytical Chemistry. 2010;**82**(15):6321-6328. DOI: 10.1021/ac100306c

[94] Liu J, Ye J, Xu C, Jiang SP, Tong Y. Kinetics of ethanol electrooxidation at Pd electrodeposited on Ti. Electrochemistry Communications. 2007;**9**(9):2334-2339. DOI: 10.1016/j.elecom.2007.06.036

[95] Yi Q, Niu F, Sun L. Fabrication of novel porous Pd particles and their electroactivity towards ethanol oxidation in alkaline media. Fuel. 2011;**90**(8): 2617-2623. DOI: 10.1016/j.fuel.2011.03.038

[96] Du H-Y, Wang C-H, Hsu H-C, Chang S-T, Chen U-S, Yen SC, et al. Controlled platinum nanoparticles uniformly dispersed on nitrogen-doped carbon nanotubes for methanol oxidation. Diamond and Related Materials. 2008;**17**(4–5):535-541. DOI: 10.1016/j.diamond.2008.01.116

[97] Kiyani R, Parnian MJ, Rowshanzamir S. Investigation of the effect of carbonaceous supports on the activity and stability of supported palladium catalysts for methanol electro-oxidation reaction. International Journal of Hydrogen Energy. 2017;**42**(36): 23070-23084. DOI: 10.1016/j.ijhydene.2017.07.113

[98] Zhang X, Zhu J, Tiwary CS, Ma Z, Huang H, Zhang J, et al. Palladium nanoparticles supported on nitrogen and sulfur dual-doped graphene as highly active electrocatalysts for formic acid and methanol oxidation. ACS Applied Materials and Interfaces. 2016;**8**(17): 10858-10865. DOI: 10.1021/acsami.6b01580

[99] Jha N, Jafri RI, Rajalakshmi N, Ramaprabhu S. Graphene-multi walled carbon nanotube hybrid electrocatalyst support material for direct methanol fuel cell. International Journal of Hydrogen Energy. 2011;**36**(12):7284-7290. DOI: 10.1016/j.ijhydene.2011.03.008

[100] Jyothirmayee Aravind SS, Imran Jafri R, Rajalakshmi N, Ramaprabhu S. Solar exfoliated graphene–carbon nanotube hybrid nano composites as efficient catalyst supports for proton exchange membrane fuel cells. Journal of Materials Chemistry. 2011;**21**(45):18199. DOI: 10.1039/c1jm13908h

[101] Pham K-C, McPhail DS, Mattevi C, Wee ATS, Chua DHC. Graphene-carbon nanotube hybrids as robust catalyst supports in proton exchange membrane fuel cells. Journal of the Electrochemical Society. 2016;**163**(3):F255-F263. DOI: 10.1149/2.0891603jes

[102] Ajeel MA, Taeib Aroua MK, Daud AW, W. M. Reactivity of carbon black diamond electrode during the electro-oxidation of remazol brilliant blue R. RSC Advances. 2016;**6**(5):3690-3699. DOI: 10.1039/C5RA21487D

Chapter 5

Progress in Cathode Materials for Methanol Fuel Cells

Joseph Parbey, Fehrs Adu-Gyamfi and Michael Gyan

Abstract

Methanol fuel cells are the most viable alternative to lithium-ion batteries for portable and other applications. The performance of methanol fuel cell depends in part on the microstructure, contact at the electrode-electrolyte interface, and oxygen reduction reactions (ORR) taking place at the cathode, which requires highly efficient cathode materials. The cathode materials have a significant impact on the performance of methanol fuel cells, making their selection and development an important field of research. This review paper provides a comprehensive overview of the progress made in cathode material selection for methanol fuel cells over the past decade. The development of different classes of cathode materials and cathode support is extensively discussed with particular emphasis on structure and electrochemical properties and performance. Also presented are research challenges and opportunities in developing new cathode materials and future trends. Finally, this review paper provides valuable insights into advancements in cathode material selection for methanol fuel cells, sheds light on hybrid composites support materials, and paves the way for further innovation in the pursuit of efficient and commercially viable methanol fuel cell technologies.

Keywords: ORR, methanol, cathode materials, electrocatalyst, fuel cells

1. Introduction

Methanol fuel cells (MFCs) are gaining popularity as an appealing alternative to traditional sources of energy due to their high energy density, low emissions, and ease of methanol management [1]. The cathode is critical in MFCs since it is where the oxygen reduction (ORR) procedure happens. To ensure effective ORR kinetics and overall cell performance, proper cathode material selection plays an essential role [2]. The efficiency of the ORR also relies on efficient catalysts at the cathode, which facilitate the process. The performance of the cathode is depended on the microstructure, high electronic conductivity, and the contact at the electrode-electrolyte interface in addition to being expected to be compatible with the cathode support [3–7]. The electrochemical oxygen reduction reaction occurring at the cathode is as follows:

$$\frac{1}{2}O_2 + 2H^+ + 2e^- = H_2O \quad (1)$$

High catalytic activities at the cathode-electrolyte interface are desired for optimum performance of the MFCs. In addition, the cathode catalyst arrangement is expected to be stable, have high selectivity, and also durable under operating conditions for a long time.

Cathode fabrication process and technique are important for microstructure optimization and improvement in the triple phase boundary (TPB), where the methanol, air, and the electrolyte react to produce the needed power for the intended application. A comprehensive review on methods of catalyst deposition of different support systems are presented by Meille [8], Di Noto et al. [7], and Singh et al. [4]. The overall performance of the cathode is mainly hindered by methanol cross-over from the anode through the membrane [9–13]. The methanol cross-over leads to the poisoning of the electrode surface and hence affect the ORR kinetics needed for power production. Further, ORR and methanol oxidation reaction (MOR) occur simultaneously at the cathode producing mixed potentials, which reduces the cell voltage and, generate additional water. Also, high cost and low durability associated with cathodes continue to be the two most important challenges restricting the commercialization of MFCs [4, 10, 14, 15].

2. MFC cathode catalyst materials

MFC cathode catalyst are mostly based on noble metals such as platinum (Pt), palladium (Pd), and ruthenium (Ru), which are expensive and hence requires efficient usage resulting in the provision of support in order to reduce cost [6, 14–17]. Furthermore, the slow reaction kinetics and surface poisoning makes it necessary to provide cathode support with improved surface area and porosity to enhance the electrochemical processes in the MFCs. Notwithstanding the setbacks, Pt-based catalysts have widely been used for ORR in both directly and supported carbon electrodes in acidic and alkaline media [11, 18, 19]. However, Pt has the highest ORR catalytic activity but develops high overpotential, which reduces the performances considerably. Also, the Pd and Ru have found wide applications in both alkaline and acidic media due to their effectiveness [6, 20].

The development of alternate catalysts as replacements for the precious metals-based is on the increase. These precious metal-based composites and non-precious metal-based cathodes are methanol and carbon monoxide (CO) tolerant with a high catalytic activity that is close to that of Pt and other precious-metal-based catalysts [5, 6, 21]. Most of these non-precious metal-based ORR catalysts are metal oxides and transition metal nitrides and sulfides and noble metals. Recently, perovskite structured metal oxides catalysts, ABO_3, which are chemically and structurally versatile, have also been proposed [13, 22]. They are reported to be ORR efficient and methanol tolerant.

3. Precious metal-based cathode materials

Precious metal-based MFC electrodes, comprising platinum (Pt), palladium (Pd), and ruthenium (Ru), meet all the basic requirements as cathodes for MFCs. The advances and significant researches into the MFCs have made precious metal-based cathode materials the most significant electrode catalyst due to their high catalytic activity, stability, and selectivity [4, 14, 20]. However, there are limitations such as methanol crossover, CO poisoning and low durability is hindering their applications.

The catalytic activity of the precious metal-based catalysts is influenced by factors such as crystallographic orientation, particle size, surface morphology, alloying effects, and catalyst-support interaction [23–25]. There have alloying of precious metals with each other to take advantages of the strengths while overcoming their weakness to enhance their activity the cathode [15, 26, 27]. For instance, Pt-Ru [26] and Pt-Pd [27] nanocomposite catalyst have been proven to improve electrocatalytic activity and long-term stability for ORR when they were used as cathodes for polymer.

3.1 Platinum-based cathode materials

Platinum (Pt) is widely regarded as the most efficient catalyst for the ORR in methanol fuel cells in MFCs. Pt-based cathodes' excellent electrochemical properties and stability make them a preferred choice for commercial applications [4, 21, 28]. Pt-based catalysts have been supported on multi-walled carbon nanotubes, carbon black, mesoporous carbon spheres, and graphene systems [4, 17, 29, 30]. Despite the importance of Pt-based cathode for MFCs and the strides made, there remain drawbacks to its utilization. The high cost and limited availability of Pt pose significant challenges for widespread application. Improvement in Pt utilization and reduction in Pt loading have been suggested to overcome these limitations.

Alloying Pt with transition metals like cobalt, nickel, or iron, synthesizing Pt-based nanocomposites, and developing advanced nanostructured catalysts have been used in addressing these concerns [11, 14, 31]. These approaches aim to enhance the catalytic activity while reducing the amount of Pt required, making the catalysts more cost-effective. Glüsen et al. [14] using these approaches reported high electrochemical performance and long-term stability for the resulting catalyst after 3000 h of operation compared to commercial Pt catalyst under the same operating conditions. Mazzapioda et al. [13] synthesized $CaTiO_{3-\delta}$ and used it as a co-catalyst with Pt/C for the cathode of MFC in order to improve the ORR for better overall performance. The presence of the $CaTiO_{3-\delta}$ introduces oxygen vacancies which serve as active sites for oxygen adsorption, and improves the ORR resulting in an increase of 40% in the maximum power density at 90°C compared to Pt/C under the same operating conditions (**Figure 1**).

Also, the electrochemical performance of the Pt-based cathode is meaningfully affected by methanol concentration (methanol crossover) and CO poisoning [4, 13, 20, 32]. Methanol crossover occurs when methanol molecules diffuse across the polymer electrolyte membrane, leading to decreased fuel efficiency and catalyst degradation. CO poisoning occurs when CO molecules adsorb onto the Pt catalyst surface, inhibiting the ORR. Development of improved catalyst formulations and exploring catalyst support materials with enhanced methanol tolerance and CO tolerance have been suggested [13, 22, 32, 33].

Further, the long-term stability and durability of Pt-based cathode materials are critical for their practical implementation in methanol fuel cells. Harsh operating conditions, such as high potentials, fuel impurities, and temperature fluctuations, can lead to catalyst degradation and performance decay [23, 25, 34, 35]. Catalyst degradation mechanisms include particle agglomeration, dissolution, and surface reconstruction. Researchers are actively working on improving the stability and durability of Pt catalysts by developing novel synthesis methods, surface modifications, and support materials to enhance catalyst resistance to degradation [35, 36].

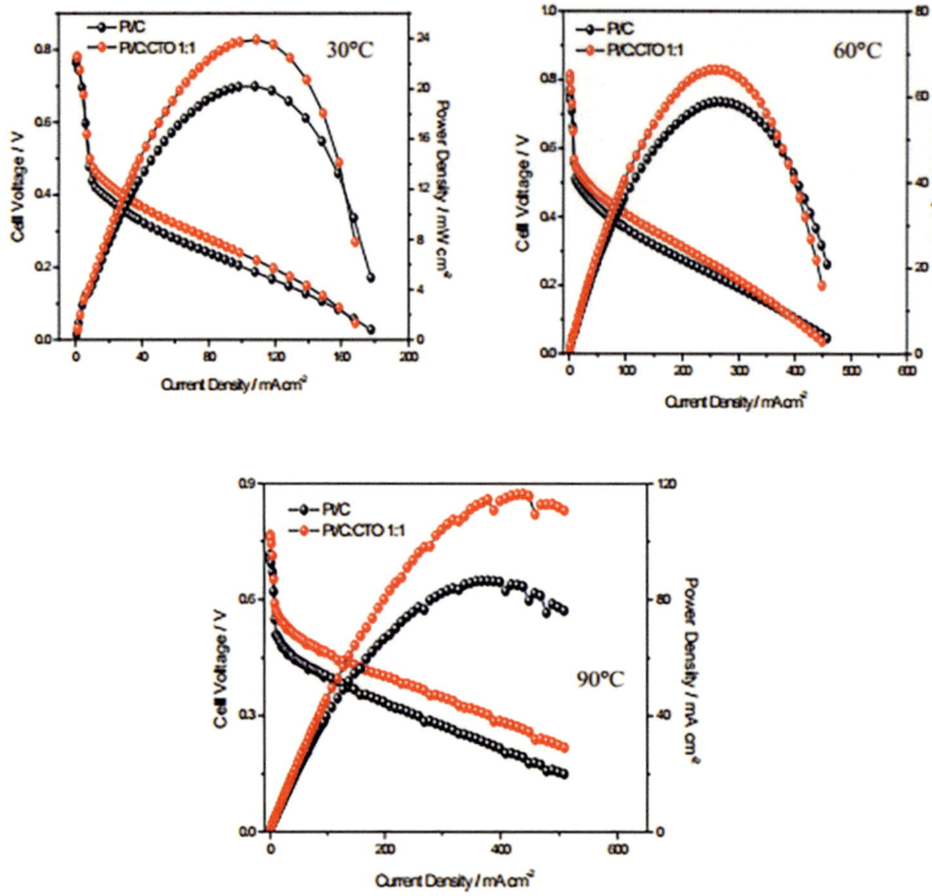

Figure 1.
Comparison of DMFC polarization and power density curves for the MEAs equipped with the bare Pt/C (black) and the composite Pt/C:CTO (red) cathode catalysts at different temperatures (30, 60 and 90°C) [13].

3.2 Palladium-based cathode materials

Palladium (Pd), a precious metal alternative to Pt-based catalyst, exhibits excellent catalytic activity for the ORR vital in methanol fuel cells [20]. Pd facilitates the electrochemical reduction of oxygen, and promotes the conversion of oxygen molecules to water. The cost-effectiveness and comparable catalytic activity to platinum make them attractive candidates for commercialization [20]. Pd-based catalysts can be utilized in direct methanol fuel cells (DMFCs) for portable electronics, transportation systems, and stationary power generation. The development of Pd-based cathode materials, along with advancements in system design and integration, can contribute to the widespread adoption of methanol fuel cells as a clean and efficient energy solution.

Although Pd is more abundant than Pt, its stability and performance can be compromised by methanol crossover and carbon monoxide poisoning. Pd has been alloyed with other metals, developing Pd-based nanomaterials with enhanced stability, and exploring novel surface modifications to improve the selectivity and durability of Pd catalysts [37]. Transition metals such as gold (Au), silver (Ag), and copper

(Cu) has been alloyed with Pd and have led to enhanced catalytic activity and stability of the composite catalyst while increasing the cathode resistance to methanol crossover and carbon monoxide poisoning [38, 39]. Li et al. [40] synthesized PdCu nanowires with improved methanol tolerance and enhanced ORR performance. Also, PdCu alloy catalysts have shown superior activity and selectivity for the ORR in the presence of methanol compared to pure Pd catalysts [41]. Such advancements demonstrate the potential of Pd-based cathode materials for methanol fuel cells.

Methanol crossover, CO poisoning, and long-term stability remain significant concerns with Pd-based electrodes, which need to be addressed. Methanol crossover can reduce fuel efficiency and degrade catalyst performance, while CO poisoning can inhibit the ORR. The stability and durability of Pd-based catalysts under harsh operating conditions require further investigation. One method suggested to overcome these drawbacks to improve Pd-based cathode performance and durability is to structure them in core-shell bifunctional catalysts [42, 43].

To improve the performance and overcome the challenges associated with Pd-based cathodes, research focused on optimizing catalyst composition, alloying strategies, and catalyst-support interactions to enhance performance should be vigorously pursued. Also, the development of novel synthesis methods, exploration of advanced characterization techniques, and investigation of new support materials will also contribute to the progress of Pd-based cathode materials [44].

3.3 Ruthenium-based cathode materials

Ruthenium (Ru) is a less commonly explored precious metal for methanol fuel cell cathodes. However, recent studies have shown that Ru-based catalysts possess notable ORR activity and exhibit superior resistance to methanol crossover and carbon monoxide poisoning [6]. The unique properties of Ru, such as its high affinity for oxygen and low reactivity towards methanol, make it an attractive alternative to Pt and Pd catalysts. For instance, Wang et al. [15] developed a Ru-based catalyst supported on a carbon nanotube matrix, which exhibited excellent catalytic performance and stability in methanol fuel cells [45]. Structural modification such as crystal phase control, nanoscale size effects, and defect engineering have played a crucial role in enhancing the ORR catalytic activity of Ru-based cathode materials [46]. These modifications optimize the active sites, surface area, and electronic properties of ruthenium catalysts, leading to improved ORR activity and stability.

Besides, Ru alloying with other transition metals has emerged as a strategy to enhance the catalytic activity and stability of Ru-based cathode materials. The alloying Ru with other transition metals enhanced the electrochemical performance since improved resistance to methanol crossover and CO poisoning are discussed, highlighting the potential of these alloy catalysts in methanol fuel cells [6]. RuSe supported on CNT cathode assembly used as cathode of DMFC for instance exhibited better electrochemical performance due to enhanced ORR resulting from the presence of Se, which improved the oxygen reduction of the RuSe/CNT compared with Pt/C [6].

To make Ru and Ru-composite-based cathode material commercially viable with improved performance and durability, further research is needed to optimize Ru-based catalysts and explore their potential in practical applications. This can be attained by exploring novel alloy compositions and understanding the underlying mechanisms to address these challenges and advance the field of ruthenium-based cathode materials for methanol fuel cells.

4. Bifunctional catalyst

To improve the selectivity and catalytic activity of MFCs, bifunctional catalysts have been investigated as potential alternatives to conventional catalysts. Bifunctional catalysts have the ability to simultaneously facilitate the ORR and MOR [19, 21]. They typically consist of a combination of metals, metal alloys, or metal oxides core and catalytically active shell, which exhibit desirable properties such as high electrochemical surface area, good electrical conductivity, and excellent stability [47]. The core material, often composed of platinum (Pt) or palladium (Pd), serves as a platform for anchoring the active sites and providing stability, while the shell material enhances the electrocatalytic activity [4, 14, 42, 43]. As cathodes for MFCs, the performance of bifunctional catalysts relies on the optimization of active sites for both the ORR as well as the promotion of intermediate species adsorption and conversion [4].

The cost associated with the use of expensive and limited supply of precious metals also makes the adoption of MFCs prohibitive. To overcome this, MFC electrodes based on bifunctional catalysts, have been designed with the core-shell consisting of a mixture of Pt or Pd with less expensive metals such as cobalt (Co), nickel (Ni) and other nanoparticles [13, 14, 30]. The resulting electrodes showed enhanced catalytic activity and stability compared to pure Pt catalyst. Zhu et al. [48], for instance, synthesized a $PtCu/CeO_2$ core-shell catalyst that exhibited excellent bifunctional catalytic activity for MOR and ORR. Also, Chen et al. [49] synthesized a PdCu/C catalyst that exhibited excellent bifunctional catalytic activity at a significantly lower cost compared to Pt-based catalysts. Various synthesis methods have been employed to fabricate efficient bifunctional catalysts for methanol fuel cell cathodes. These methods include wet chemical synthesis, solvothermal/hydrothermal methods, electrodeposition, and physical deposition techniques [13, 50]. The choice of synthesis method depends on factors such as catalyst composition, morphology, and desired catalytic properties [51]. Controlling the catalyst structure at the nanoscale level is crucial for achieving enhanced catalytic activity [52].

Bifunctional catalysts have demonstrated significant improvements in the performance of methanol fuel cell cathodes. The integration of both the ORR and MOR functionalities in a single catalyst enhances the overall cell efficiency, reduces the reliance on separate catalysts for each reaction, and mitigates issues related to catalyst poisoning [47]. These have been attained through the use of multi-element catalysts. For instance, Pt, the most commonly used ORR catalyst, has been alloyed with transition metals such as gold (Au), ruthenium (Ru), iridium (Ir), and others to form a bifunctional catalyst in order to improve the catalytic activities and MOR tolerance [5, 19]. Zhang et al. (2021) synthesized a PtAuCu/C catalyst with significantly enhanced MOR and ORR activities compared to pure Pt catalysts. The alloy catalyst showed improved tolerance to CO poisoning, making it a promising candidate for methanol fuel cells. Pérez et al. [19] synthesized Pt/Cr/Ru on carbon and studied their MOR and ORR. The Pt/Cr/Ru substrate prepared by simultaneously depositing them on the carbon electrode showsthe best activity for ORR and is less affected by methanol. Baglio et al. [53] fabricated a bifunctional Pt-Fe/C cathode and reported enhanced catalytic activity, methanol tolerance, and enhanced ORR. Also gaining prominence for ORR is a Pt-free cobalt spinel oxides $M_XCo_{3-x}O_4$ where M is transitional metal made up of Fe, Co, Ni, and Mn [5]. Usually, carbon, graphene, and other conducting surfaces are attached to or used as support, thus forming the shell for these spinel oxides [54, 55]. These compounds have comparative catalytic activity in addition to their low cost due to their abundance and environmental

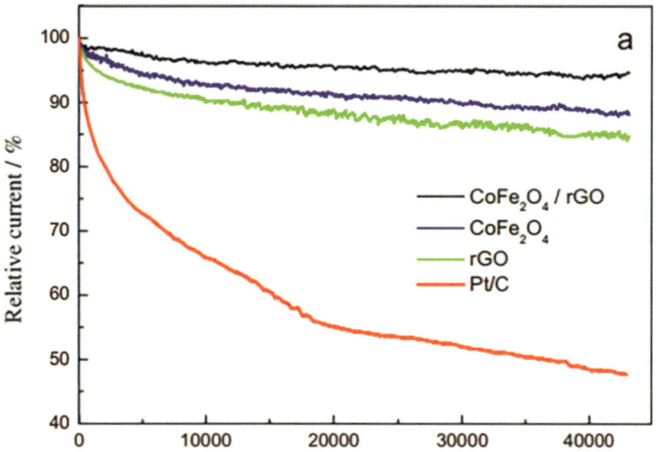

Figure 2.
Current-time (i-t) chronoamperometric responses for the ORR on $CoFe_2O_4$ (mixed with AB), rGO, $CoFe_2O_4$/rGO nanohybrid and commercial Pt/C in O_2-saturated 0.1 M KOH at 0.4 V (vs. Ag/AgCl) [5].

friendliness. Bian et al. [5] reported a highly porous nanohybrid cobalt spinel oxide $CoFe_2O_4$ on graphene electrode structure with large surface area, good adhesion, excellent pore distribution provided a large TPB for ORR, and stability compared with Pt/C operating under the same conditions (**Figure 2**).

Bifunctional catalysts have shown great potential for enhancing the performance and stability of methanol fuel cell cathodes. Their ability to simultaneously catalyze the ORR and the MOR offer advantages such as improved cell efficiency, reduced catalyst reliance, and enhanced tolerance to methanol crossover. Despite the progress made in bifunctional catalyst research, several challenges remain. The optimization of catalyst composition, design, structure, and synthesis techniques to achieve synergistic effects between the ORR and the MOR is essential for performance optimization. Also, enhancing the stability of bifunctional catalysts under the harsh operating conditions of MFCs is also a critical area that needs much attention and focus. Moreover, the development of cost-effective and scalable synthesis methods for large-scale production of bifunctional catalysts is necessary for their commercial viability.

5. Transition metal nitrides and sulfides (TMNS)

Material properties and electrode characteristics are important in the performance of MFCs. To improve performance, Transition metal nitrides and sulfides (TMNS) have attracted significant attention and are proposed as potential catalyst materials for methanol fuel cell cathodes. TMNS are usually Pt-free electrocatalysts consisting of transition metals such as Fe, Mn, Co, etc., and non-metal groups, mainly N and C [5, 7, 50] . TMNS and its electrodes have desirable properties such as high conductivity, abundant active sites, and good electrocatalytic activity required for effective ORR at the cathode [13, 56, 57]. In addition, TMNS electrodes are durable and methanol tolerant [57]. TMNS cathodes have been developed for ORR commercially and reported to be inactive to MOR, and have desirable properties such as high conductivity, abundant sites for catalytic activity, and good electrocatalytic activity required for ORR effective ORR at the cathode [13, 56, 57]. Vecchio et al. [58] studied the intrinsic activity of commercial Fe-N-C electrocatalysts towards ORR and MOR inactivity by rotating disk electrode electrochemical characterization technique. They reported MOR tolerance and high activity for ORR. Fajardo et al. [50] also synthesized and studied the ORR activity of nitrogen and sulfur-doped Co_3O_4 nanoparticles on graphene for MFC. The resulting electrode had lower area-specific resistance and, higher power and current densities than commercial Pt-based catalysts. The electrode assembly exhibited good stability and durability after a long run.

WN, MoN, CoS, and MoS_2, TMNS used as MFC cathode, have also been studied and reported to efficiently enhance ORR kinetics resulting in improved MFC performance due to their material characteristics and microstructure of the electrode [32, 57]. Tungsten nitride (WN) and molybdenum nitride (MoN), for instance, have been reported to exhibit excellent electrical conductivity and high chemical stability [59]. Besides, they excellently possess catalytic activity towards ORR due to the presence of nitrogen vacancies that provide active sites for the oxygen reduction process. Further, cobalt sulfide (CoS) and molybdenum sulfide (MoS_2), possess a layered structure with abundant edge sites that promote catalytic activity [60]. They also exhibit good stability and high electrical conductivity, making them attractive candidates for methanol fuel cell cathodes. The specific material properties desired for enhanced catalytic performance inform the choice of method for TMNS MFC cathode fabrication [60]. Several methods such as chemical vapor deposition, solvothermal synthesis, hydrothermal synthesis, and electrodeposition have been adopted in order to control the morphology, composition, and particle size of the TMNS [61].

Generally, the performance of the Pt-metal-free cathodes is greatly influenced by methanol concentration, catalyst loading, and operation temperature [43]. High methanol concentrations cause high rates of methanol crossover, hence reduced performance [57]. However, for Pt-metal-free cathodes, the effect of high methanol crossover due to higher methanol concentration is countered by increasing the catalyst loading and temperature, where the performance improves remarkably since the ORR kinetics and MOR tolerances rise. The power density of Fe-C-N catalyzed cathode in MFC increased from about 33 to 45 mW cm^{-2} when the catalyst loading was 2–6 mg/cm^2 [58], and also the power density increased with temperature (60–90°C), catalyst loading (2–6 mg/cm^2) and methanol concentration (1–10 M) with the open circuit potential fairly constant throughout.

Although the use of TMNS for MFC cathodes have seen progressive development, there remain several challenges. Optimized catalyst composition, morphology, and surface structure are vital to further enhance the catalytic activity and stability of TMNS. Additionally, understanding the reaction mechanisms and the role of specific active sites needs to be thoroughly explored. Also, the integration of these catalysts into practical fuel cell systems is needed and requires addressing issues such as scalability, cost-effectiveness, and long-term stability. To address these, further optimization and tailoring of the TMNS cathode microstructure through the exploration of electrode fabrication mechanisms is recommended to further improve the overall cell performance. Also, further exploration of synthesis methods, catalyst design, and mechanistic studies will enable the realization of efficient and cost-effective methanol fuel cells.

6. Support materials for MFC cathode catalyst materials

The cathode support material plays a crucial role in enhancing the catalytic activity, stability, and durability of catalysts for ORR in methanol fuel cells [62].

Carbon-based materials, metal oxides, and hybrid composites have demonstrated promising results in enhancing catalytic activity and durability of cathode catalysts. The physical properties of the catalyst support, such as porosity, pore size distribution, adhesion of catalyst, and others, affect the performance of the cell [22, 63, 64] . Nanostructured cathode support assembly improves the porosity and provides more active sites for oxygen species adsorption, dissociation, partial reduction, and the combination of oxygen species with oxygen vacancy on the electrode surface [65, 66]. Carbon-based catalyst support systems are widely used. CNFs and CNTs, carbon-based nanostructured cathode support for MFCs cathode, are reported to improve ORR, methanol tolerance, and stability of MFC cathodes [3, 29, 67]. Also, metal oxides [22, 64, 68] and hybrid composites [69–71] have also been used and reported to improve ORR and long-term stability of the cathode support systems.

6.1 Carbon-based support materials

Carbon-based materials have been extensively investigated as cathode support materials for methanol fuel cells due to their excellent electrical conductivity, chemical stability, and high surface area. Additionally, the high corrosion resistance, uniform particle size distribution, strong adhesion of catalyst particles, and uniform dispersion of catalyst particles on support make carbon-based materials more desirable [5, 29, 67, 72]. They support and stabilize the catalyst nanoparticles, facilitate ORR, and improve the overall performance of MFCs. Carbon black, Carbon nanotubes (CNTs), Graphene, Mesoporous carbon, and Carbon nanofibers (CNFs) are the main carbon-based materials for MFC cathode support [17, 29, 62, 73]. Carbon-supported cathodes become unstable leading to loss of catalyst activity during long operation. The loss of activity and hence reduced ORR is due to oxidation of carbon, which splits the catalyst particles leading to reduced cell performance [67]. Also, the dissolution, sintering, and agglomeration of Pt leads to the corrosion of carbon support materials leading to the low durability of the cell [15].

Carbon black is a commonly used carbon-based support material due to its high electrical conductivity, large surface area, and well-established commercial availability [17, 72]. It provides a three-dimensional network structure that enhances the dispersion and stability of catalyst nanoparticles. Surface modifications, such as acid treatment or nitrogen doping, have been employed to modify the surface properties of carbon black, leading to improved catalytic activity and methanol tolerance [74, 75]. For instance, nitrogen doping introduces active sites and modifies the electronic structure of carbon, enhancing the catalytic activity and stability of the catalyst [7, 76]. The carbon/nitrogen hybrid cathode support composites can be synthesized through various methods such as pyrolysis of carbon precursors in the presence of nitrogen-containing compounds. Carbon/nitrogen-doped carbon composites exhibit improved ORR activity, enhanced methanol tolerance, and increased durability [77]. Zhang et al. [78] reported that nitrogen-doped carbon nanofiber composites exhibited superior catalytic performance and stability in methanol fuel cells.

To improve methanol tolerance and cathode stability during the operation of MFCs, carbon nanotubes (CNT) with highly conductive pathways for electron transport and effectively promoting the ORR kinetics have been fabricated and used as cathode support [79]. The unique structure of CNTs also allows for strong interaction with catalyst nanoparticles, enhancing catalyst dispersion and stability. Surface functionalization of CNTs with functional groups, such as carboxyl or hydroxyl, further improves their compatibility with catalyst nanoparticles and enhances catalytic activity [80, 81]. Luo et al. reported that functionalized CNTs-supported Pt catalysts exhibited superior methanol tolerance and stability compared to traditional carbon black supports. Also, nitrogen-doped CNT supported Pt-Ru [43].

Carbon nanofibers (CNF) are three-dimensional porous structures synthesized through catalytic growth or electrospinning techniques [3, 27]. CNF offers unique physical properties such as porous structure for enhanced mass transport and electrochemical performance, and large surface area for easy accessibility of reactants to catalysts. CNFs, just like any carbon-based support material, have high electrical conductivity and have been proven to enhance methanol tolerance, and excellent durability [82]. For example, Zhang et al. reported that Pt-loaded CNFs exhibited higher catalytic activity and better stability in methanol fuel cells compared to carbon black supports [83]. Similar reported enhanced catalytic activity and long-term stability have been reported by several authors [27].

Graphene is a two-dimensional carbon material with high electrical conductivity, large surface area, high porosity, and excellent mechanical strength [5, 84]. It has emerged as a promising cathode support material for methanol fuel cells due to its physicochemical properties. Graphene-based support materials can be synthesized through chemical vapor deposition and exfoliation techniques [16, 67, 85, 86]. Graphene supports exhibit improved catalytic activity and stability due to enhanced charge transfer and efficient dispersion of catalyst nanoparticles. Wang et al. [45] demonstrated that graphene-supported Pt catalysts exhibited higher ORR activity and improved methanol tolerance compared to traditional carbon supports [45]. Also, Brian et al. [5] reported enhanced ORR catalytic activity and stability when a bimetal $CoFe_2O_4$ was supported on graphene. Moreover, a survey on the performance of polymer electrolyte fuel cells (PEMFCs) showed that Pt and Pt-alloyed catalysts and non-precious metal catalysts on graphene and graphene-doped cathode supports have superior catalytic activity and stability for ORR resulting in improved cell performancc [26, 67].

Mesoporous carbon is porous structured carbon fabricated in such a manner in order to optimize the morphology, porosity, and intimate adhesion of the catalyst to the surface for enhanced ORR activity [29, 62]. Mesoporous carbon support has been reported to have increased the power density by about 30% compared to normal carbon structured support due to the high specific surface area, porosity, and adhesion leading to extended TPB and enhanced ORR [29].

6.2 Metal oxide support materials

Metal oxides such as titanium dioxide (TiO_2), tin dioxide (SnO_2), and cerium oxide (CeO_2) have been investigated and used as cathode catalyst support materials in MFCs [64, 73, 87, 88]. They offer unique properties, such as high thermal stability, corrosion resistance, and strong interaction with catalyst nanoparticles. Metal oxide support systems provide stable platforms for catalyst anchoring, improving the catalytic activity, methanol tolerance, and durability of the catalyst [23, 64, 88]. However, the ability to control the metal oxide morphology, optimize the metal-support interaction, and mitigate of catalyst poisoning and methanol crossover effects, coupled with the stability and long-term durability of metal oxide supports under harsh operating conditions, limits its practical implementation. Several methods, such as surface modifications and other engineering techniques, have been proposed to enhance their electrocatalytic properties and subsequent performance when used as cathode supports [23].

Titanium dioxide (TiO_2) is a widely explored metal oxide support material in MFCs due to its exceptional chemical stability, high surface area, and low cost [64, 89–91]. TiO_2 provides a stable and conductive platform for catalyst immobilization, allowing for efficient charge transfer and improved catalytic activity [68, 88, 91]. They also show long-term durability, electrochemical oxidation, and corrosion resistance [23, 88, 91]. Ioroi et al. [91] used Ti_4O_7 as support for Pt (Pt/Ti_4O_7) and reported good catalytic activity for ORR and has the potential to be used as corrosion resistant cathode. Huang et al. [88] also reported the exhibition of good electrochemical performance under full cell operation and corrosion resistance and enhanced stability of the cathode system due to the presence of TiO_2 support. Surface modifications, such as doping with metals or nitrogen, have been employed to enhance the electrocatalytic properties of TiO_2 supports [89]. For instance, Hassen et al. [92] reported that nitrogen-doped carbon-supported TiO_2 nanofiber catalysts exhibited enhanced ORR activity and methanol tolerance. A mixed metal oxide of tantalum (Ta) modified TiO_2 prepared by the sulfite complex route was used as a support for Pd catalyst and evaluated for stability and durability as an ORR catalyst [23]. The presence of the Ta modified the structural and surface properties leading to an increase the oxygen deficiency and improved morphology, which led to high ORR activity, stability, and durability compared with other catalysts.

Tin dioxide (SnO_2) has emerged as a promising cathode support material due to its high electrical conductivity, chemical stability, and strong interaction with catalyst nanoparticles [64, 87]. SnO_2 supports promote better catalyst dispersion and stability, enhancing the catalytic performance of methanol oxidation. Zhang et al. [87] synthesized mesoporous SnO_2 as cathode support for Pt to investigate the electrochemical performance, stability, and durability in PEMFCs. The Pt/SnO_2 cathode assembly presented a good surface area, good electrochemical activity for ORR, and long-term durability and stability compared to Pt/C. The electrochemical surface area loss was only 50% compared with 90% for the Pt/C. Surface modifications, such as alloying with other metals or metal oxides, have been employed to improve the catalytic activity and methanol tolerance of SnO_2-based supports [78]. For example, Wang et al. [93] demonstrated that Pt-SnO_2 nanocomposites exhibited enhanced catalytic performance and durability in methanol fuel cells [93].

Cerium oxide (CeO_2), also known as ceria, with excellent oxygen storage capacity, redox properties, and high surface area, has attracted attention as a cathode support material for MFCs. CeO_2 supports facilitate the oxygen supply and storage during the ORR, improving the catalytic activity and methanol tolerance of the catalyst. To improve the catalytic properties of ceria support, it has either been doped with metals or defects introduced into its structure in order to modify the catalyst surface to tailor it for ORR [94]. Liu et al. [95] reported that metal-doped CeO_2-supported Pt catalysts exhibited enhanced catalytic activity and stability in methanol fuel cells.

6.3 Hybrid composite support materials

Hybrid composite materials, formed by combining carbon-based materials with metal oxides or other nanomaterials, have gained attention as promising cathode support materials. Carbon/metal oxide composites, carbon/polymer composites, and carbon/nitrogen-doped carbon composites have been synthesized and used as cathode support to enhance catalytic, improve stability, and synergistic effects [64, 96]. They have enhanced electrical conductivity, improved catalyst dispersion, and increased catalytic activity.

Carbon/metal oxide composites cathode support takes advantage of the physiochemical properties of carbon and metal oxide to form a superior catalyst support to over the individual limitations associated with these individual materials. These composites can be synthesized using various methods such as sol-gel, chemical vapor deposition, or hydrothermal approaches [96]. Carbon/metal oxide composites exhibit improved catalyst dispersion, enhanced catalytic activity, and stability compared to individual components [93]. For example, Zhu et al. [97] reported that carbon/manganese oxide composites showed enhanced catalytic activity and durability for the oxygen reduction reaction in methanol fuel cells. Nanostructured ZrO_2/nitrogen-doped graphene nanosheets were synthesized and used as support for Pt exhibited higher electrochemical surface area, durability, and better ORR attributable to the unique structure and chemical interactions between the support and the catalyst [96].

Also, polymers such as polyvinyl alcohol (PVA), polyacrylonitrile (PAN), or polypyrrole (PPY) have been used in combination with carbon-based materials to form composite supports [98]. The integration of polymer into carbon-based materials improved the mechanical strength, stability, and facile processability, while carbon impacts electrical conductivity and a large surface area for improved ORR. Carbon/polymer composites exhibit enhanced catalyst dispersion, improved methanol tolerance, and excellent durability [98]. For instance, PAN-based carbon composites are reported to exhibit superior performance and stability in methanol fuel cells [99, 100]. Also, polymers have solely been used as support materials for Pt, Ru, and Pd catalysts for ORR in polymer electrolyte fuel cells. For instance, polypyrrole, due to its electrical conductivity and environmental stability among others, have been used as support for Pt and Pd with corresponding improvement in ORR activities [98, 101]. The methods of the PPY-catalyst preparations greatly influence the ORR activity.

Like all other cathode support materials, the hybrid composites face challenges such as long-term stability, methanol crossover, and catalyst poisoning effects although they have unique properties that enhance the catalytic activity for MFCs. Careful understanding of the fundamental principles governing the catalytic performance of hybrid composite cathode support materials, and further optimization of the microstructure of these hybrid composite supports through continued research to pave the way for efficient and durable methanol fuel cells.

7. Summary and future outlook

Notwithstanding the significant advances in materials selection for MFC cathodes, the use of Pt and Pt/composite electrodes still dominates as the material of choice. Sluggish oxygen reduction reaction, due to methanol crossover from the anode and inherently slow reaction kinetics, has been identified as the main problem associated with the cathode of MFCs coupled with the associated catalyst cost. Efforts should be geared towards the development of new catalysts aimed at improved reaction kinetics through the use of new cathode and cathode-support materials. Deliberate catalyst design, synthesis techniques optimization, and mechanistic understanding of the catalytic activity are crucial for performance optimization and commercialization of MFCs.

Furthermore, the development of low-cost but effective ORR catalysts, with high methanol tolerance and CO tolerance or high selectivity for ORR is highly desirable. The development of low-cost catalyst alternatives with comparable performance is crucial for commercial viability. However, challenges such as stability and durability must be addressed to ensure their widespread adoption.

DOI: http://dx.doi.org/10.5772/intechopen.1003869

In conclusion, creating novel cathode materials is crucial to tackling the world's energy issues. Despite the inherent difficulties and limitations, the discipline is set to make tremendous progress. To realize the full potential of cathode materials, researchers are increasingly relying on multidisciplinary approaches, better characterization techniques, and novel synthesis methods. Future advances in cathode material research promises to transform the landscape of energy technology, allowing a cleaner and more sustainable energy future as demand for efficient and sustainable energy storage and conversion solutions continues to climb.

Author details

Joseph Parbey[1*], Fehrs Adu-Gyamfi[2] and Michael Gyan[3]

1 Industrial Chemistry Unit, Department of Chemistry, University of Cape Coast, Cape Coast, Ghana

2 Department of Mechanical Engineering, Koforidua Technical University, Koforidua, Ghana

3 Department of Physics Education, University of Education, Winneba, Ghana

*Address all correspondence to: joseph.parbey@ucc.edu.gh

References

[1] Ahmed M, Dincer I. A review on methanol crossover in direct methanol fuel cells: Challenges and achievements. International Journal of Energy Research. 2011;**35**(14):1213-1228

[2] de Sá MH, Moreira CS, Pinto AMFR, Oliveira VB. Recent advances in the development of nanocatalysts for direct methanol fuel cells. Energies. 2022;**15**(17):6355

[3] Subianto S et al. Electrospun nanobers for low-temperature proton exchange membrane fuel cells. In: Cavaliiere S, editor. Electrospinning for Advanced Energy and Environmental Applications. Boca Raton: CRC Press; 2016. pp. 29-60

[4] Singh RN, Awasthi R, Sharma CS. Review: An overview of recent development of platinum-based cathode materials for direct methanol fuel cells. International Journal of Electrochemical Science. 2014;**9**(10):5607-5639

[5] Bian W, Yang Z, Strasser P, Yang R. A $CoFe_2O_4$/graphene nanohybrid as an efficient bi-functional electrocatalyst for oxygen reduction and oxygen evolution. Journal of Power Sources. 2014;**250**:196-203

[6] Jeng K-T, Hsu N-Y, Chien C-C. Synthesis and evaluation of carbon nanotube-supported RuSe catalyst for direct methanol fuel cell cathode. International Journal of Hydrogen Energy. 2011;**36**(6):3997-4006

[7] Di Noto V et al. Hierarchical metal–[carbon nitride shell/carbon core] electrocatalysts: A promising new general approach to tackle the ORR bottleneck in low-temperature fuel cells. ACS Catalysis. 2022;**12**(19): 12291-12301

[8] Meille V. Review on methods to deposit catalysts on structured surfaces. Applied Catalysis A: General. 2006;**315**:1-17

[9] Jörissen L, Gogel V, Kerres J, Garche J. New membranes for direct methanol fuel cells. Journal of Power Sources. 2002;**105**(2):267-273

[10] Aricò AS, Srinivasan S, Antonucci V. DMFCs: From fundamental aspects to technology development. Fuel Cells. 2001;**2**(1):133-161

[11] Antolini E, Lopes T, Gonzalez ER. An overview of platinum-based catalysts as methanol-resistant oxygen reduction materials for direct methanol fuel cells. Journal of Alloys and Compounds. 2008;**461**(1):253-262

[12] Shaari N, Kamarudin SK, Bahru R, Osman SH, Ishak NTIM. Progress and challenges: Review for direct liquid fuel cell. International Journal of Energy Research. 2021;**5**(45):6644-6688

[13] Mazzapioda L, Lo Vecchio C, Aricò AS, Navarra MA, Baglio V. Performance improvement in direct methanol fuel cells by using $CaTiO_{3-\delta}$ additive at the cathode. Catalysts. 2019;**9**(12):1017

[14] Glüsen A et al. Dealloyed PtNi-core–shell nanocatalysts enable significant lowering of Pt electrode content in direct methanol fuel cells. ACS Catalysis. 2019;**9**(5):3764-3772

[15] Wang Y-J, Wilkinson DP, Zhang J. Noncarbon support materials for polymer electrolyte membrane fuel cell electrocatalysts. Chemical Reviews. 2011;**111**(12):7625-7651

[16] Zuo Y, Sheng W, Tao W, Li Z. Direct methanol fuel cells system–A

review of dual-role electrocatalysts for oxygen reduction and methanol oxidation. Journal of Materials Science & Technology. 2022;**114**:29-41

[17] Ramli ZAC, Kamarudin SK. Platinum-based catalysts on various carbon supports and conducting polymers for direct methanol fuel cell applications: A review. Nanoscale Research Letters. 2018;**13**(1):410

[18] Ensafi AA, Jafari-Asl M, Rezaei B. A new strategy for the synthesis of 3-D Pt nanoparticles on reduced graphene oxide through surface functionalization, application for methanol oxidation and oxygen reduction. Electrochimica Acta. 2014;**130**:397-405

[19] Pérez G, Pastor E, Zinola CF. A novel Pt/Cr/Ru/C cathode catalyst for direct methanol fuel cells (DMFC) with simultaneous methanol tolerance and oxygen promotion. International Journal of Hydrogen Energy. 2009;**34**(23):9523-9530

[20] Álvarez GF, Mamlouk M, Scott K. An investigation of palladium oxygen reduction catalysts for the direct methanol fuel cell. International Journal of Electrochemistry. 2011;**2011**:684535

[21] Dang Long Q, Phuoc Huu L. Recent advances in Pt-based binary and ternary alloy electrocatalysts for direct methanol fuel cells. In: Lindiwe Eudora K, editor. Electrocatalysis and Electrocatalysts for a Cleaner Environment. Rijeka: IntechOpen; 2021. p. Ch. 1

[22] Li L, Tan S, Salvatore KL, Wong SS. Nanoscale perovskites as catalysts and supports for direct methanol fuel cells. Chemistry – A European Journal. 2019;**25**(33):7779-7797

[23] D'Urso C, Bonura G, Aricò AS. Synthesis and physical-chemical characterization of nanocrystalline Ta modified TiO_2 as potential support of electrocatalysts for fuel cells and electrolyzers. International Journal of Hydrogen Energy. 2017;**42**(46):28011-28021

[24] Cai Y et al. Recent advances in Ni-based catalysts for CH_4-CO_2 reforming (2013-2023). Atmosphere. 2023;**14**(9):1323

[25] Ke Y et al. An overview of noncarbon support materials for membrane electrode assemblies in direct methanol fuel cells: Fundamental and applications. Materials & Design. 2023;**233**:112261

[26] Woo S, Lee J, Park S-K, Kim H, Chung TD, Piao Y. Enhanced electrocatalysis of PtRu onto graphene separated by Vulcan carbon spacer. Journal of Power Sources. 2013;**222**:261-266

[27] Lv J-J et al. Simple synthesis of platinum–palladium nanoflowers on reduced graphene oxide and their enhanced catalytic activity for oxygen reduction reaction. Journal of Power Sources. 2014;**269**:136-143

[28] Abdullah N, Rahman S, Mohd Zainoodin A, Aslfattahi N. Comparative study for electrochemical and single-cell performance of a novel MXene-supported platinum–ruthenium catalyst for direct methanol fuel cell application. Journal of Electroanalytical Chemistry. 2022;**925**:116884

[29] Bruno MM, Viva FA, Petruccelli MA, Corti HR. Platinum supported on mesoporous carbon as cathode catalyst for direct methanol fuel cells. Journal of Power Sources. 2015;**278**:458-463

[30] Gan L, Heggen M, Rudi S, Strasser P. Core–shell compositional fine structures of dealloyed PtxNi1–x

nanoparticles and their impact on oxygen reduction catalysis. Nano Letters. 2012;**12**(10):5423-5430

[31] Toda T, Igarashi H, Uchida H, Watanabe M. Enhancement of the electroreduction of oxygen on Pt alloys with Fe, Ni, and Co. Journal of the Electrochemical Society. 1999;**146**(10):3750

[32] Karim NA, Kamarudin SK, Loh KS. Performance of a novel non-platinum cathode catalyst for direct methanol fuel cells. Energy Conversion and Management. 2017;**145**:293-307

[33] Antolini E. Effect of structural characteristics of binary palladium-cobalt fuel cell catalysts on the activity for oxygen reduction. ChemPlusChem. 2014;**79**(6):765-775

[34] Zeng Y et al. Pt nanoparticles on atomic-metal-rich carbon for heavy-duty fuel cell catalysts: Durability enhancement and degradation behavior in membrane electrode assemblies. ACS Catalysis. 2023;**13**:11871-11882

[35] Bai P, Wang P, Mu J, Xie Z, Du C, Su Y. Toward the long-term stability of cobalt benzoate confined highly dispersed PtCo alloy supported on a nitrogen-doped carbon nanosheet/Fe3C nanoparticle hybrid as a multifunctional catalyst for zinc-air batteries. ACS Applied Materials & Interfaces. 2023;**15**(29):35117-35127

[36] Chen S et al. Review of SOFC cathode performance enhancement by surface modifications: Recent advances and future directions. Energy & Fuels. 2023;**37**(5):3470-3487

[37] Li H, Li G. Novel palladium-based nanomaterials for multifunctional ORR/OER/HER electrocatalysis. Journal of Materials Chemistry A. 2023;**11**(17):9383-9400

[38] Ashraf M, Ahmad MS, Inomata Y, Ullah N, Tahir MN, Kida T. Transition metal nanoparticles as nanocatalysts for Suzuki, heck and Sonogashira cross-coupling reactions. Coordination Chemistry Reviews. 2023;**476**:214928

[39] Abhishek N, Verma A, Singh A, Kumar T. Metal-conducting polymer hybrid composites: A promising platform for electrochemical sensing. Inorganic Chemistry Communications. 2023;**157**:111334

[40] Li C-J, Shan G-C, Guo C-X, Ma R-G. Design strategies of Pd-based electrocatalysts for efficient oxygen reduction. Rare Metals. 2023;**42**(6):1778-1799

[41] Zhang P et al. Alloy as advanced catalysts for electrocatalysis: From materials design to applications. Chinese Chemical Letters. 2023:109073

[42] Symillidis A, Georgiadou S, Lin W-F. Conductive core/shell polymer nanofibres as anode materials for direct ethanol fuel cells. Advanced Sensor and Energy Materials. 2023;**2**(3):100070

[43] Luo L et al. Ultrafine Core@ Shell Cu1Au1@ Cu1Pd3 nanodots synergized with 3D porous N-doped graphene nanosheets as a high-performance multifunctional electrocatalyst. ACS Nano. 2023;**17**(3):2992-3006

[44] Zhang X, Wang J, Zhao Y. Enhancement mechanism of Pt/Pd-based catalysts for oxygen reduction reaction. Nanomaterials. 2023;**13**(7):1275

[45] Wang B, Bi L, Zhao XS. Liquid-phase synthesis of SrCo0.9Nb0.1O3-δ cathode material for proton-conducting solid oxide fuel cells. Ceramics International. 2018;**44**(5):5139-5144

[46] Li J-C, Hou P-X, Cheng M, Liu C, Cheng H-M, Shao M. Carbon nanotube encapsulated in nitrogen and phosphorus co-doped carbon as a bifunctional electrocatalyst for oxygen reduction and evolution reactions. Carbon. 2018;**139**:156-163

[47] Belenov S et al. The PtM/C (M = Co, Ni, Cu, Ru) electrocatalysts: Their synthesis, structure, activity in the oxygen reduction and methanol oxidation reactions, and durability. Catalysts. 2023;**13**(2):243

[48] Zhu Z, Qian J, Wang Z, Dang J, Liu W. High-performance anode-supported solid oxide fuel cells based on nickel-based cathode and Ba(Zr0.1Ce0.7Y0.2)O3−δ electrolyte. Journal of Alloys and Compounds. 2013;**581**:832-835

[49] Chen Y, Liu H, Zhuang L, Wei Y, Wang H. Hydrogen permeability through Nd5.5W0.35Mo0.5Nb0.15O11.25-δ mixed protonic-electronic conducting membrane. Journal of Membrane Science. 2019;**579**:33-39

[50] Fajardo S, Ocón P, Rodríguez JL, Pastor E. Co supported on N and S dual-doped reduced graphene oxide as highly active oxygen-reduction catalyst for direct ethanol fuel cells. Chemical Engineering Journal. 2023;**461**:142053

[51] Jaramillo-Páez C, Navío JA, Hidalgo MC. Effect of synthesis pH on the physicochemical properties of a synthesized Bi2WO6 and the type of substrate chosen, in assessing its photo-catalytic activities. Arabian Journal of Chemistry. 2020;**13**(1):431-443

[52] Mitchell S, Qin R, Zheng N, Pérez-Ramírez J. Nanoscale engineering of catalytic materials for sustainable technologies. Nature Nanotechnology. 2020;**16**:1-11

[53] Baglio V, Stassi A, Di Blasi A, D'Urso C, Antonucci V, Aricò AS. Investigation of bimetallic Pt–M/C as DMFC cathode catalysts. Electrochimica Acta. 2007;**53**(3):1360-1364

[54] Huang X, Qi X, Boey F, Zhang H. Graphene-based composites. Chemical Society Reviews. 2012;**41**(2):666-686. DOI: 10.1039/C1CS15078B

[55] Yuan XZ, Qu W, Zhang X, Yao P, Fahlman J. Spinel NixCo2-xO4 as a bifunctional air electrode for zinc air batteries. ECS Transactions. 2013;**45**(29):105

[56] Tang S, Zhang Z, Xiang J, Yang X, Shen X, Song F. Recent advances in transition metal nitrides for hydrogen electrocatalysis in alkaline media: From catalyst design to application (in English). *Frontiers in Chemistry,* Mini Review. 2022;**10**

[57] Meng Z et al. Transition metal nitrides for electrocatalytic application: Progress and rational design (in eng). Nanomaterials (Basel). 2022;**12**(15):2660

[58] Lo Vecchio C et al. Commercial platinum group metal-free cathodic electrocatalysts for highly performed direct methanol fuel cell applications. Journal of Power Sources. 2019;**437**:226948

[59] Jaf ZN, Altarawneh M, Miran HA, Jiang Z-T. Geometries, electronic properties and stability of molybdenum and tungsten nitrides low-index surfaces. Materials Research Express. 2018;**5**(12):126402

[60] Cárdenas-Lizana F, Lamey D, Kiwi-Minsker L, Keane MA. Molybdenum nitrides: A study of synthesis variables and catalytic performance in acetylene hydrogenation. Journal of Materials Science. 2018;**53**(9):6707-6718

[61] Luo Q, Lu C, Liu L, Zhu M. A review on the synthesis of transition metal nitride nanostructures and their energy related applications. Green Energy & Environment. 2023;**8**(2):406-437

[62] Sharma S, Pollet BG. Support materials for PEMFC and DMFC electrocatalysts—A review. Journal of Power Sources. 2012;**208**:96-119

[63] Liu Y, Ji C, Gu W, Jorne J, Gasteiger HA. Effects of catalyst carbon support on proton conduction and cathode performance in PEM fuel cells. Journal of the Electrochemical Society. 2011;**158**(6):B614

[64] Lv H, Mu S. Nano-ceramic support materials for low temperature fuel cell catalysts. Nanoscale. 2014;**6**(10):5063-5074. DOI: 10.1039/C4NR00402G

[65] Parbey J, Wang Q, Yu G, Zhang X, Li T, Andersson M. Progress in the use of electrospun nanofiber electrodes for solid oxide fuel cells: A review. Reviews in Chemical Engineering. 2020;**36**(8):879-931

[66] Cavaliere S, Cavaliere S, editors. Electrospinning for Advanced Energy and Environmental Applications-CRC Press. Boca Raton: CRC Press; 2016. p. 302

[67] Zhou X, Qiao J, Yang L, Zhang J. A review of graphene-based nanostructural materials for both catalyst supports and metal-free catalysts in PEM fuel cell oxygen reduction reactions. Advanced Energy Materials. 2014;**4**(8):1301523-1301523

[68] Shanmugam S, Gedanken A. Synthesis and electrochemical oxygen reduction of platinum nanoparticles supported on mesoporous TiO_2. The Journal of Physical Chemistry C. 2009;**113**(43):18707-18712

[69] Choi J-H, Park K-W, Lee H-K, Kim Y-M, Lee J-S, Sung Y-E. Nano-composite of PtRu alloy electrocatalyst and electronically conducting polymer for use as the anode in a direct methanol fuel cell. Electrochimica Acta. 2003;**48**(19):2781-2789

[70] Choi J-H et al. A polyaniline supported PtRu nanocomposite anode and a Pd-impregnated nanocomposite Nafion membrane for DMFCs. Solid State Ionics. 2005;**176**(39):3031-3034

[71] Wu H et al. Polypyrrole nanowire networks as anodic micro-porous layer for passive direct methanol fuel cells. Electrochimica Acta. 2014;**141**:1-5

[72] Wang G, Sun G, Wang Q, Wang S, Sun H, Xin Q. Effect of carbon black additive in Pt black cathode catalyst layer on direct methanol fuel cell performance. International Journal of Hydrogen Energy. 2010;**35**(20):11245-11253

[73] Du L, Shao Y, Sun J, Yin G, Liu J, Wang Y. Advanced catalyst supports for PEM fuel cell cathodes. Nano Energy. 2016;**29**:314-322

[74] Chen T-W et al. Recent developments in carbon-based nanocomposites for fuel cell applications: A review. Molecules. 2022;**27**(3):761

[75] Lefèvre M, Proietti E, Jaouen F, Dodelet J-P. Iron-based catalysts with improved oxygen reduction activity in polymer electrolyte fuel cells. Science. 2009;**324**(5923):71-74

[76] Di Noto V, Negro E, Gliubizzi R, Gross S, Maccato C, Pace G. Pt and Ni carbon nitride electrocatalysts for the oxygen reduction reaction. Journal of the Electrochemical Society. 2007;**154**(8):B745

[77] Chen Q, Zhang Z, Zhang R, Hu M, Shi L, Yao Z. Recent progress of non-Pt catalysts for oxygen reduction reaction in fuel cells. Processes. 2023;**11**(2):361

[78] Zhang J et al. A new highly active and CO_2-stable perovskite-type cathode material for solid oxide fuel cells developed from A- and B-site cation synergy. Journal of Power Sources. 2020;**457**:227995

[79] Forootan Fard H, Khodaverdi M, Pourfayaz F, Ahmadi MH. Application of N-doped carbon nanotube-supported Pt-Ru as electrocatalyst layer in passive direct methanol fuel cell. International Journal of Hydrogen Energy. 2020;**45**(46):25307-25316

[80] Alosime EM. A review on surface functionalization of carbon nanotubes: Methods and applications. Discover Nano. 2023;**18**(1):12

[81] Chen J, Chen Q, Ma Q. Influence of surface functionalization via chemical oxidation on the properties of carbon nanotubes. Journal of Colloid and Interface Science. 2012;**370**:32-38

[82] Huang H, Wang X. Recent progress on carbon-based support materials for electrocatalysts of direct methanol fuel cells. Journal of Materials Chemistry A. 2014;**2**(18):6266-6291

[83] Zhang D, Yang Z, Li P, Pang M, Xue Q. Flexible self-powered high-performance ammonia sensor based on Au-decorated $MoSe_2$ nanoflowers driven by single layer MoS_2-flake piezoelectric nanogenerator. Nano Energy. 2019;**65**:103974

[84] Geim AK, Novoselov KS. The rise of graphene. Nature Materials. 2007;**6**(3):183-191

[85] Ambrosi A, Chua CK, Bonanni A, Pumera M. Lithium aluminum hydride as reducing agent for chemically reduced graphene oxides. Chemistry of Materials. 2012;**24**(12):2292-2298

[86] Cho SH, Yang HN, Lee DC, Park SH, Kim WJ. Electrochemical properties of Pt/graphene intercalated by carbon black and its application in polymer electrolyte membrane fuel cell. Journal of Power Sources. 2013;**225**:200-206

[87] Zhang P, Huang S-Y, Popov BN. Mesoporous tin oxide as an oxidation-resistant catalyst support for proton exchange membrane fuel cells. Journal of the Electrochemical Society. 2010;**157**(8):B1163

[88] Huang S-Y, Ganesan P, Zhang P, Popov B. Development of novel metal oxide supported Pt catalysts for polymer electrolyte membrane and unitized regenerative fuel cells applications. ECS Transactions. 2009;**25**(1):1893

[89] Sun S, Zhang G, Sun X, Cai M, Ruthkosky M. Highly stable and active Pt/Nb-TiO_2 carbon-free electrocatalyst for proton exchange membrane fuel cells. Journal of Nanotechnology. 2012;**2012**:389505

[90] Tominaka S. Facile synthesis of nanostructured reduced titanium oxides using borohydride toward the creation of oxide-based fuel cell electrodes. Chemical Communications. 2012;**48**(64):7949-7951. DOI: 10.1039/C2CC33532H

[91] Ioroi T, Senoh H, Yamazaki S-I, Siroma Z, Fujiwara N, Yasuda K. Stability of corrosion-resistant Magnéli-phase Ti_4O_7-supported PEMFC catalysts at high potentials. Journal of the Electrochemical Society. 2008;**155**(4):B321

[92] Hassen D et al. Nitrogen-doped carbon-embedded TiO_2 nanofibers as promising oxygen reduction reaction

electrocatalysts. Journal of Power Sources. 2016;**330**:292-303

[93] Wang W, Xu X, Zhou W, Shao Z. Recent progress in metal-organic frameworks for applications in electrocatalytic and photocatalytic water splitting. Advance Science (Weinh). 2017;**4**(4):1600371

[94] Ma J, Tao Z, Kou H, Fronzi M, Bi L. Evaluating the effect of Pr-doping on the performance of strontium-doped lanthanum ferrite cathodes for protonic SOFCs. Ceramics International. 2020;**46**(3):4000-4005

[95] Liu Y et al. Physical properties of (SrBa)1-xPrx(CuTi)0.2Fe0.8O3-δ (x = 0-1.0) and its application in H-SOFCs. Solid State Ionics. 2020;**348**:115279

[96] Cheng N et al. High stability and activity of Pt electrocatalyst on atomic layer deposited metal oxide/nitrogen-doped graphene hybrid support. International Journal of Hydrogen Energy. 2014;**39**(28):15967-15974

[97] Zhu A et al. Evaluation of SrSc0.175Nb0.025Co0.8O3-δ perovskite as a cathode for proton-conducting solid oxide fuel cells: The possibility of in situ creating protonic conductivity and electrochemical performance. Electrochimica Acta. 2018;**259**:559-565

[98] Jeyabharathi C, Venkateshkumar P, Mathiyarasu J, Phani KLN. Carbon-supported palladium–polypyrrole nanocomposite for oxygen reduction and its tolerance to methanol. Journal of the Electrochemical Society. 2010;**157**(11):B1740

[99] Berber MR, Althubiti NA, Alrowaili ZA, Rosa F, Iranzo A. Boosting the catalytic efficiency of platinum nanoparticles supported on pristine carbon nanotubes: Synergistic effects of conducting polymers. Fuel. 2021;**306**:121681

[100] Bezerra CWB et al. A review of Fe–N/C and Co–N/C catalysts for the oxygen reduction reaction. Electrochimica Acta. 2008;**53**(15):4937-4951

[101] Góral-Kurbiel M, Drelinkiewicz A, Kosydar R, Gurgul J, Dembińska B, Kulesza PJ. The effect of Nafion ionomer on electroactivity of palladium–polypyrrole catalysts for oxygen reduction reaction. Journal of Solid State Electrochemistry. 2014;**18**(3):639-653